2020

VISION

Health in the 21st Century

Institute of Medicine 25th Anniversary Symposium

INSTITUTE OF MEDICINE

NATIONAL ACADEMY PRESS
Washington, D.C. 1996

National Academy Press • 2101 Constitution Avenue, N.W. • Washington, DC 20418

NOTICE: The project that is the subject of this report was approved by the Institute of Medicine's Council.

This report has been reviewed by a group other than the authors according to procedures approved by a Report Review Committee consisting of members of the National Academy of Sciences, the National Academy of Engineering, and the Institute of Medicine.

The Institute of Medicine was chartered in 1970 by the National Academy of Sciences to enlist distinguished members of the appropriate professions in the examination of policy matters pertaining to the health of the public. In this, the Institute acts under both the Academy's 1863 congressional charter responsibility to be an adviser to the federal government and its own initiative in identifying issues of medical care, research, and education. Dr. Kenneth I. Shine is president of the Institute of Medicine.

The Institute gratefully acknowledges the generous support of the following organizations toward its 25th anniversary activities: Albert Einstein College of Medicine of Yeshiva University; Carnegie Corporation of New York; Columbia University College of Physicians and Surgeons; The Commonwealth Fund; Kaiser-Permanente; Memorial Sloan-Kettering Cancer Center; The Mount Sinai Hospital/Mount Sinai School of Medicine; The New York Hospital; The Pew Charitable Trusts; Times Mirror Company; University of California, Irvine; University of Florida Health Science Center; University of Medicine and Dentistry of New Jersey; University of Rochester Medical Center; University of Southern California; and Yale University School of Medicine. *Corporate benefactor:* Bristol-Myers Squibb. *Corporate patrons:* The Hoffmann-La Roche Foundation; Merck & Co., Inc.; and Pfizer, Inc. *Corporate sponsors:* Amgen, Inc.; Eli Lilly and Company; Marion Merrell Dow, Inc.; Procter and Gamble; Sandoz Pharmaceuticals Corporation; Schering-Plough Corporation; G.D. Searle and Company; and The Upjohn Company.

International Standard Book Number: 0-309-05488-5
Library of Congress Catalog Card Number: 96-68279

Additional copies of this publication are available from: National Academy Press, Lock Box 285, 2101 Constitution Avenue, N.W., Washington, DC 20055.

Call (800) 624-6242 or (202) 334-3313 in the (Washington metropolitan area), or visit the NAP on-line bookstore at **http://www.nap.edu/nap/bookstore/**.

Call (202) 334-2352 for more information on the other activities of the Institute of Medicine, or visit the IOM home page at **http://www.nas.edu/iom/**.

The serpent has been a symbol of long life, healing, and knowledge among almost all cultures and religions since the beginning of recorded history. The serpent adopted as a logotype by the Institute of Medicine is a relief carving from ancient Greece, now held by the Staatlichemuseen in Berlin.

COVER: Plate #753 by Robert Sperry. Stoneware, white slip over black glaze, 1987. Courtesy of the National Museum of American Art, Smithsonian Institution; gift of the James Renwick Alliance. This and other of Mr. Sperry's works were shown at the National Academy of Sciences through Arts in the Academy, a public service program of the National Academy of Sciences.

Cover design by Linda Humphrey.

INSTITUTE OF MEDICINE 25TH ANNIVERSARY SYMPOSIUM
PLANNING COMMITTEE

KENNETH I. SHINE, M.D. (*Chair*), president, Institute of Medicine, Washington, D.C.

JORDAN J. COHEN, M.D., president, Association of American Medical Colleges, Washington, D.C.

SR. ROSEMARY DONLEY, S.C., executive vice president, The Catholic University of America

JOHN M. EISENBERG, M.D., chairman and physician-in-chief, Department of Medicine, Georgetown University Medical Center

RUTH S. HANFT, Ph.D., consultant, Alexandria, Virginia

BARBARA C. HANSEN, Ph.D., director, Obesity and Diabetes Research Center, and Professor, Department of Physiology, School of Medicine, University of Maryland at Baltimore

MICHAEL M. E. JOHNS, M.D., vice president and dean of the Medical Faculty, The Johns Hopkins University School of Medicine

ELAINE L. LARSON, Ph.D., dean, Georgetown University School of Nursing

CLAUDE LENFANT, M.D., director, National Heart, Lung and Blood Institute, National Institutes of Health, Bethesda, Maryland

FITZHUGH MULLAN, M.D., director, Bureau of Health Professions, and assistant surgeon general, U.S. Public Health Service, Rockville, Maryland

ROBERT F. MURRAY, JR., M.D., professor of pediatrics, medicine, and genetics, Howard University College of Medicine

BARBARA STARFIELD, M.D., University Distinguished Service Professor, Division of Health Policy, The Johns Hopkins University School of Hygiene and Public Health

DONALD E. WILSON, M.D., dean, School of Medicine, University of Maryland at Baltimore

Staff

KAREN HEIN, M.D., executive officer

JANA H. SURDI, director, Office of Council and Membership Services

JOAN SIEBER, administrative assistant, Office of Council and Membership Services

DON TILLER, administrative assistant, Division of Health Care Services (until February 1996)

Preface

After about a decade of planning, consultation, and negotiation, 30 people received a charter on December 17, 1970, to establish the Institute of Medicine (IOM). As part of the National Academy of Sciences (NAS), the IOM's mission is to advance scientific knowledge and the health and well-being of all people of this nation and the world consistent with the roles conferred upon it by its congressional authority.

That congressional authority extends back to the 1863 NAS charter authorized by the Congress and signed by President Abraham Lincoln, in which the Academy was asked to advise the government on issues related to science and the arts and to do so without a fee. Thus, all of our work is done by committees of experts who work without honoraria, pro bono, with the support and assistance of an outstanding staff. The Institute is greatly indebted to its many members and others who have worked on these committees over the years. This list of people, through whose service the nation and world have benefited, is long and distinguished. We deeply appreciate their efforts and commitment.

The Institute accomplishes its mission by providing objective, timely, and authoritative information to government, the health professions, and the public through both its elected membership and through its access to people with the insight and expertise needed to tackle major issues of the day. The results of IOM's work are exemplified by the variety of reports and other activities produced over the years and the impact we have had on society.

As part of the ongoing celebration of the IOM's 25th anniversary over the last year, we have already touched upon our past accomplishments, both at the 25th Anniversary Annual Meeting and in the publication *For the Public Good: Highlights from the Institute of Medicine, 1970–1995.*

Therefore, we thought it fitting to dedicate the 25th Anniversary Symposium to a look ahead at the next 25 years. The title of the symposium, "2020 Vision," was not selected to satisfy our members who are ophthalmologists, but in recognition of the fact that IOM will then be 50 years old.

What challenges will we face in the years to come, and what forces will affect our lives, our health, and our health care system in 2020? If we can take a clue from the past, the challenges are likely to be significant and, at least to some degree, unpredictable. In December 1970, at the time of IOM's founding, Hodgkin's disease was essentially uniformly fatal, childhood leukemias were largely fatal, bone marrow transplants were a highly experimental activity, and many kinds of organ transplants were only hopeful dreams in the minds of researchers. Thanks to technological advances and research breakthroughs, that picture has changed considerably in the intervening years. However, we also have witnessed the emergence of HIV/AIDS and the reemergence of old enemies such as tuberculosis— serious reminders that we should not take too much comfort in the battles we've won, as we still have far to go.

Therefore, we look to the future grounded in a rich past, and confident that whatever lies ahead, the Institute of Medicine will be there to play its part in finding solutions, answering questions, and guiding policymakers in the effort to improve the public health, both in the United States and worldwide.

Kenneth I. Shine
President, Institute of Medicine

Acknowledgments

As with all endeavors such as this, "2020 Vision: Health in the 21st Century"—both the symposium itself and this volume containing its proceedings—would not have been possible without the efforts and energy of a large group of dedicated people. These include the members of the symposium planning committee: Jordan J. Cohen, Sr. Rosemary Donley, John M. Eisenberg, Ruth S. Hanft, Barbara C. Hansen, Michael M. E. Johns, Elaine L. Larson, Claude Lenfant, Fitzhugh Mullan, Robert F. Murray, Jr., Barbara Starfield, and Donald E. Wilson; the Institute of Medicine staff who helped organize and carry off the event: Joan Sieber, Jana Surdi, and Don Tiller; IOM's executive officer, Karen Hein; our distinguished speakers: Baruch S. Blumberg, Robert M. Carey, Lincoln C. Chen, Don E. Detmer, John M. Eisenberg, Richard G. A. Feachem, Jeff Goldsmith, Paul F. Griner, Michael M. E. Johns, Lawrence S. Lewin, Edward H. O'Neil, the Honorable John E. Porter, the Honorable Donna E. Shalala, and Donald E. Wilson.

In addition, we appreciate the work of Claudia Carl and Mike Edington from the Institute's Reports and Information Office, Dawn Eichenlaub and Sally Stanfield from the National Academy Press, and our graphic designer, Linda Humphrey, for their help in the production and publication of this volume.

Kenneth I. Shine
President, Institute of Medicine

Contents

2020
VISION

Health in the 21st Century

Overview and Introduction

Donald E. Wilson, M.D., M.A.C.P.
Dean, School of Medicine, University of Maryland at Baltimore

During the last three years, no subject has generated more interest, more debate, or more activity than our nation's health and health care delivery system. Industry began to see a third of its expenditures used to provide health benefits to its employees. Consumers grew alarmed by health insurance premiums that escalated rapidly to thousands of dollars a year and locked them into their current position due to a lack of portability of coverage for preexisting conditions.

The nation saw its health bills consume 14 percent of the gross national product, or $900 billion. However, there is an interesting paradox:

- most Americans have access to unparalleled state-of-the-art health care and are satisfied with the treatment that they receive;
- people from all over the world come to the United States in search of the most advanced medical procedures and technology;
- our medical education system produces some of the world's finest physicians and scientists; and
- the United States is a world-recognized leader in biomedical research. Babies weighing less than one pound now survive and grow to become healthy children, and we have made remarkable progress toward unlocking the mysteries of the human genome.

At the same time, however, 40 million Americans are without health insurance, and another 30 million Americans do not have adequate health insurance. We have seen the reemergence of preventable diseases such as polio and measles, and the continuation of devastating but preventable disabilities caused by lead poisoning, the return of tuberculosis, and a veritable explosion of sexually transmitted diseases.

Consumers became confused or ambivalent: thankful for, but expecting and demanding, that high-technology subspecialty care be available when they needed it, yet also concerned about their ability to afford and obtain general care that attended to their health as well as their diseases. Conventional wisdom indicated that something needed to be done.

The year 1994 was one of great health care nonreform. Although national legislation was not enacted, states began to seek ways to deal seriously with the issues—primarily the cost, but also the quality, of care.

Medical schools were told that the country did not need or want the current mix of physicians. Indeed, in some instances, public institutions were told that their continued funding was dependent upon a certain percentage of their graduates choosing primary care specialties. More important, in a climate increasingly intolerant of regulation but enormously supportive of cost reduction, industry took charge.

In 1995 we have witnessed dramatic changes in the delivery of health care and in the payment for health care services. Managed care and capitation have expanded greatly. The State of Maryland, which was in the early stages of managed care penetration 2 years ago, has moved to among the top four regions at the present time. Certain types of physician specialists have found much less demand for their services and a corresponding decrease in their compensation. Academic health centers are scrambling to maintain their enterprises and fulfill their missions, and are confused about what is a successful strategy.

Most recently, some experts and policymakers have begun to question whether the changes are too rapid; whether they may have long-term negative effects on our educational, research, and discovery abilities; and whether the changes that have occurred are indeed providing affordable, high-quality health care. Will we be able to solve the problem of people who are uninsured? Will the poor have access to appropriate health care?

As is usually the case, before a problem can be solved it must be defined. What are the needs of our population, and what will be the major problems in the future? What new programs can best be devised to meet these needs? What information is necessary to assess these needs, and how do we communicate this information effectively? Can we develop a reasonable payment system that adequately compensates health care providers and, at the same time, motivates them to use resources judiciously, engage in health care rather than disease treatment, and keep the welfare and well-being of patients at the forefront of their activities? In the current frenzy of competition, will we be able to sustain our premier educational and research institutions in order to support the training of providers and investigators and the development of new clinical and technological discoveries? Should there be a significant downsizing of the academic enterprise, and if so, how should we accomplish this? How do we utilize most effectively the talents of an increasingly diversified, multidisciplinary health work force to best provide affordable health care to the patients?

This 25th Anniversary Symposium raises many of these issues and challenges us to move aggressively to develop solutions. We may also have an opportunity to look into a variety of crystal balls and sneak a peak at the rapidly approaching 21st century.

Opening Address

Honorable John E. Porter (R-IL), J.D.
U.S. House of Representatives

L et me begin by congratulating the Institute on 25 years of service to our country and to the world. The subject of this symposium is health in the 21st century. I am not a futurist; I am an appropriator. We tend to think in terms of one year at a time, sometimes seven years at a time, but no longer. Looking ahead to the year 2020 seriously challenges my ability to envision the future, but I will try.

I thought I would focus on four major elements of the issue: first, the structure of the health care system and who pays for it; second, the demographics that will place demands on it; third, health care personnel demand and supply; and last, the biomedical science knowledge base on which the health system operates.

I expect to highlight issues and raise questions rather than offer prescriptions. My focus is principally on the biomedical research enterprise and how it fits into the health care system of the future.

As many of you know, the health of the National Institutes of Health (NIH) has been one of my principal concerns in Congress. In addition, I feel I must include in my remarks a brief discussion of the larger budgetary environment in which we are currently operating, beginning with some observations about the health care system that we pay for.

The history of the health care system has been characterized by periodic cycles of momentous change. It seems that we can identify important events every 15 years or so: The enactment of the Medicare system in 1965, the movement toward the diagnosis-related group (DRG) payment system in the 1980s, and now the transformation of Medicare and Medicaid that is being considered in the Balanced Budget Act.

There is no reason to think that this trend will not continue. So I suspect that by the year 2020, we will already have seen at least one additional round

4

of changes that will lead to fundamental differences in the division and financing of health care.

One trend I would highlight is the movement toward increasing consumer choice and competition in health care, which I expect to accelerate. It is a common thread running through much of the economic restructuring occurring in the U.S. economy and across the world. It has taken place in technology-based industries such as telephone, television, and personal communications, and it was seen earlier in the service industries such as air, rail, and truck transportation.

The movement toward choice and competition is influencing private-sector health care. Sooner or later, publicly financed programs will feel its effect as well. This competition will lead to an increased emphasis on trying to determine what works in health care in a cost-effective way—the kind of information provided by the Agency for Health Care Policy and Research (AHCPR) and others who are interested in outcomes and comparative analysis.

Increased competition may even change the solutions to health care problems that are developed by the biomedical community. Will researchers be as quick to find expensive technological fixes if they know that their application depends on a finding of cost-effectiveness as well as of efficacy?

I am also convinced that the information revolution is bound to come to health care sooner or later. Up until now, it seems that health care has lagged behind in its adaptation of information systems technology, but its possibilities for reducing costs through this technology are easy to envision—from standardization of medical claims payments, to automated patient records, to tele-medicine. All of these will soon be with us.

The question of who pays for health care is likely to change as well. We see fewer people with the protection of retirement programs that support health care. Solvency problems with Medicare dictate increasing constraints on that system's ability to shoulder costs. The decreasing willingness on the part of policymakers to have Medicare reimburse the direct and indirect expenses of graduate medical education is one example of this.

Whatever the outcome of the current budget debate, Medicaid is sure to see constraints on its future growth as well, with no obvious solutions to the problem of uncompensated care. We have also seen private industry lead the federal government in its efforts to contain medical costs for its employees and to cut back on benefits.

As a by-product of constrained resources, we will continue to see rationing of health care. The development of medical technology is outstripping the resources needed to provide medical services. This rationing is most often conflicted and based on economics—benefits go to those who can pay for them. Sometimes the rationing is done by policymakers and is explicit, as in the case of Oregon's health care plan for low-income people.

This potentially troubling picture of financing health care in the future is not brightened by looking at the second area I will discuss—demographics. I confront this area regularly in my work with the Foreign Operations Appropriations Subcommittee. The demographic trends worldwide present acute challenges, and we cannot be complacent in thinking that the United States is immune from these same problems.

The one obvious example is aging and the increasing proportion of the population that will be over 65 in the years to come. You need only look at projections of the incidence of Alzheimer's disease, and the consequent long-term-care costs, to understand that demographics may be destiny as far as the health care system is concerned.

Problems with the income and educational levels of the population are also bound to affect the provision of health care and access to it, people's knowledge base about healthy life styles, and other issues.

Another vital component of the health care system of the future, and the third area I would like to address, is personnel supply and demand. The perceived glut of medical specialists and shortage of primary care providers has been a hot topic recently. In addition, recent reports like that from the Pew Health Professions Commission, among others, raise the more general supply issue—that we simply have too many doctors. If this perceived imbalance is corrected, either through reimbursement policies or through the laws of supply and demand, it may have the beneficial effect of reducing health care costs. It could be a significant problem, however, in generating the skilled personnel needed for biomedical research and for the environment of teaching hospitals in which some of this research is conducted.

In addition, the issue of whether the Ph.D. training system should undergo changes has been raised. Recommendations have been made to modify the content of training programs. There is also the overall supply issue: Much contested, but at least legitimate, questions have been raised as to whether the unemployment and underemployment of Ph.D.s suggests that the supply of Ph.D. students should also be adjusted.

My fourth principal area of discussion is the health knowledge base supported through biomedical research. The basic funding structure of biomedical research has not changed substantially since the post-World War II period. It remains a closely knit coalition of academia, industry, and government funders, although the participation of private sector biotechnology and pharmaceutical firms has obviously increased over the past decade. It is possible that this web of interdependent partners will be reexamined in terms of its structure and financing, spurred in large part by changes in health care and health personnel.

Questions may be raised about whether the distribution of public resources among the various actors in this partnership matches national priorities. The economic benefits accruing to the various partners—in the form of licenses, patents, product sales, and the like—may be examined as part of the scrutiny of the allocation of public resources.

The structure of the biomedical coalition, which leaves the federal government the responsibility for funding basic research and leaves the principal responsibility for funding applied research to others, may also be reexamined.

Although it is a question I am reluctant to raise, others may ask if the nation will have enough resources to maintain the current model of NIH, one that spreads money widely for basic research, with the assumption that a few successes will finance or justify the general investment, or whether funding constraints will force the system into a more directed approach to research, with funding allocated only for particular purposes. Will funding pressures cause a shift toward financing a smaller community of researchers rather than supporting the research infrastructure of 200 separate universities?

Will there be a rethinking of the size of intramural research at NIH and throughout government, with the determination either (a) that in a period of constraint, it fulfills a mission that cannot be supported as efficiently outside or (b) that it is no longer needed, given the maturation of the extramural community? Will NIH, either on its own or at the urging of others, develop new mechanisms for funding research?

I see the NIH model of distributing research funding through peer review of innovative ideas submitted by investigators around the country as one that should be adopted more widely throughout the government. However, the demands of the 21st century may dictate new mechanisms to enhance this successful system.

Finally, I cannot conclude without some comment on the nation's budgetary situation and its impact on the health programs in which I have a deep interest. As an appropriator, I have a special concern about the financial burden that health care entitlements will place on federal revenues and the resulting squeeze on funding available for important discretionary health programs—among them, NIH; the Centers for Disease Control and Prevention; health services programs such as family planning, community health centers, and Ryan White AIDS services; and veterans' health care.

I hope you are as concerned as I am by the long-term budgetary situation we face. By 2012, 17 years from now, unless substantial changes are made now, entitlements and interest on the public debt will consume the entire federal budget. By 2030, 35 years from now, Medicare, Medicaid, Social Security, and federal employee retirement programs alone will consume all of the tax revenues collected by the federal government. This means that without significant changes in the growth of entitlements, there will be no money at all to support discretionary spending of any sort.

Although the current budget negotiations are an important effort in reining in the cost of entitlements, we need to remember that previous budget balancing efforts have tended to be quite ephemeral. If you recall, we went through major budget balancing exercises in 1981, 1982, 1986, 1988, 1990, and 1993. We were assured by the Congressional Budget Office in every case that the budget would be in balance in a matter of years. It never has been.

This is not in any way intended to denigrate our current efforts, the success of which I feel is absolutely essential, but it is intended to serve as a context for what I view as the single overriding concern facing discretionary health accounts—an absolute drying up of resources necessary to sustain them. Unless we take legislative action that will actually work to reduce the cost of entitlement programs, the discretionary health accounts will face increasingly bitter competition for resources that grow more scarce each year. As resources for services and biomedical research decline, this will have tremendously negative consequences for the health of the American people.

DISCUSSION*

DR. SHINE: What is your view of the future of AHCPR?

REPRESENTATIVE PORTER: It is probably instructive to examine what happens when an agency has the courage to look at subjects honestly and give its views. In reference to AHCPR, it is clear to me that the agency offended a group in the medical community that then asked its members of Congress to cut AHCPR's budget.

I am a great fan of AHCPR, but I realized that a very strong effort was being made to cut its budget and I thought I might be able to head this off by making some cuts of my own. In the mark-up we made a fairly large cut—given the importance of AHCPR's endeavors—to the point that we decreased discretionary spending in my subcommittee by $9 billion, out of $70 billion in discretionary funds.

I thought that would head off further cuts; it did not. Further cuts occurred both in the subcommittee mark-up and on the floor of the House of Representatives. The budget as it stands in the House bill has been cut drastically—I think about in half. I do not expect this to be the final result if we can get an overall agreement by Congress and the White House on spending.

Until recently, I had been fairly sanguine that we would get that agreement, but I am becoming increasingly less sanguine that that will happen. I said from the very beginning that the best scenario for NIH and other health care funding is to reach agreements early between the White House and Congress to keep high priorities for areas in which our country leads the world and cannot afford to reduce spending—indeed, must increase it. Unfortunately, that has not occurred. In the current budgetary context we may see a continuing resolution for a year's funding for the Departments of Labor, Health and Human Services, and Education, perhaps at 75 or 80 percent of last year's level, laid on the president's desk while Congress adjourns. That would be an absolute disaster.

I am doing everything possible to influence the leaders of Congress to place a high priority on health care financing. This is all being done among a few people at the very top—our budgetary chairmen, Representative John Kasich and Senator Pete Domenici. Both budgets cut NIH and other health care matters very heavily, and I am worried about where we are and about the lack of progress or the lack of forthcomingness on the part of the White House to find some common ground and get our budgetary situation finalized.

PARTICIPANT: How will we obtain more appropriate levels of funding for research?

---

*Because of technical difficulties, the participants' questions had to be reconstructed; therefore, they are not verbatim.

REPRESENTATIVE PORTER: There needs to be a much stronger message from medical organizations such as the Institute of Medicine that send policymakers in a direction in which they are perhaps reluctant to go. I do not say this in any way as a criticism. I say simply that I think you have far more credibility than you realize, and if you can mount an effort to bring issues like this not before the Congress but before the American people, and get the attention of the media—which, in today's world, is everything—the chances of changing policy are great.

I have said to Harold Varmus, who is doing a magnificent job at NIH, that he has to become the Carl Sagan of biomedical research and popularize all of the wonderful things that NIH is funding and that are occurring in biomedical research because people are intuitively supportive. They need to understand what is happening, and how exciting it is, and the possibilities for its development.

The same kinds of things must happen with respect to policy issues. We have to raise the consciousness of the whole country about the meaning of what we are doing or not doing and capture people's understanding a bit better. We talk about this among ourselves, but we must realize that, in a big society such as ours, decisions are really made broadly, and if we cannot get broad interest in things, it is very difficult to break through and achieve change.

PARTICIPANT: I am concerned that ideology has compromised support of international health and family planning. Is this not short sighted?

REPRESENTATIVE PORTER: I share your concern. First, across this land there is a feeling that we in the federal government spend huge amounts of our resources overseas. Even in districts that are as educated and well informed as mine, you will find that people believe that we spend anywhere from 5 to 15 percent of our federal resources on foreign aid of one type or another. The fact is that all types of foreign aid—economic and military, bilateral and multilateral—amount to less than 1 percent of the federal budget, and this figure is decreasing: It has gone down about 40 percent over the last 5 or 6 years. That function of government has contributed more to deficit reduction than any other. This year, foreign aid has decreased $2 billion. Very frankly, I worry greatly that we are withdrawing our resources at a time when the United States is in the best position to influence the future of life on this planet and to emphasize the value of freedom, democracy, and family planning in international health.

I believe that there is clearly an attempt in this Congress, based on the abortion issue, to undermine all money for family planning. A large group in the House of Representatives is simply opposed to family planning. Its members will say that they are opposed to abortion, the fact is that the attack is on family planning itself, not just on abortion. I worry that the United States is going to pull back on family planning and send a message to others who are interested in this area that they do not need to commit resources either.

PARTICIPANT: How do you think the physician oversupply will affect the job market?

REPRESENTATIVE PORTER: When you have a free society and people can go into different professions or businesses, you may find that in a health care system in which some competition seems to be developing, the overall attractiveness from an economic standpoint of being a physician decreases and the willingness of people who are trained as physicians to dedicate themselves and spend inordinate amounts of time helping others may also decrease. You may see people going into other professions. There also may be greater opportunities for women in health than in a lot of other areas, when they run up against "glass ceilings" and have nowhere to go with their talent, health may be the area to which they can turn.

PARTICIPANT: Isn't it important to clarify what we actually mean by reductions in Medicare spending? I don't think we're talking about actual reductions but reduced projected spending.

REPRESENTATIVE PORTER: Neither do I, and neither does anybody else who has looked at it. In fact, under the Republican budget, spending per Medicare beneficiary will rise from $4,800 currently to $7,100 by the year 2002. We are not spending less on Medicare. We are spending substantially more, but we are not spending at the historic rate of about 10 or 10.5 percent.

This is very interesting, because the president himself has made a proposal to restrain the rate of growth in Medicare. Hillary Clinton said this at the time she offered her program. The difference between our proposal and the president's is very small, about $130 per year per beneficiary.

I guess you are asking me to make a political comment, which I am willing to make on the president's proposal. I think it is totally irresponsible. No one who looks at the federal budget and understands what is happening with entitlements—and Medicare is the largest one, while Medicaid is becoming the next largest—can fail to understand that we have to restrain the rate of increase in these programs or we are simply going to lose everything in terms of discretionary spending.

NIH—in all that it does, including public health—is at risk if we do not gain control over entitlement spending. We have to do it; nobody in this country who understands anything about the federal budget doubts that for a minute; yet the president is playing games with it. It is time for him to sit down at the table and come to an agreement; otherwise, our future is going to be very, very bleak.

PARTICIPANT: It seems that the support for children is evaporating in this country. What do you see as the short- and long-term consequences of Congress' actions?

REPRESENTATIVE PORTER: Well, I certainly agree with the premise that although we have largely solved senior poverty in our country, children are the people most at risk. Obviously, they do not have political clout, as the seniors do. They do not vote at all. Even the youngest who are eligible do not vote in any great numbers, so they do not get the weight they should receive in our deliberations; and you are right about that.

On the other hand, if we do not gain control over federal spending, we are destroying their opportunity for any kind of an economic life in our society.

I will repeat something that I have said for the last 13 years, and the numbers have gotten worse rather than better: If you are a young person entering the work force in America today, you are being handed a bill by your government for about $187,000 of extra taxes that is your share of the interest on our current national debt. even if we bring the budget into balance, this will increase by $1.2 trillion more, so your costs will be still higher. That is just the interest for an average young child. For the children and grandchildren of people in this room, the bill will be $10,000 a year, perhaps higher. We have got to put an end to this and get control over federal spending. By the way, we should not be cutting taxes at such a time. That is nonsense, and I have been against it from the very beginning.

We must look at what we are doing to our children. Are we providing them with a sound education? Is the Chicago public school system acceptable? No. Is the Medicare/Medicaid system acceptable for people in poverty? No. Maybe the idea of letting the states take primary responsibility—they already administer the Medicaid program—is not a good idea. Maybe that is something we should not do.

Clearly here, as well, with costs increasing at a rate of about 10 percent per year, we must restrain the rate of increasing costs. We have to put into place plans that the states can develop to emphasize cost-effectiveness and have resources spent wisely for needed health care reform.

Some states will do a terrible job. Others will do a magnificent job—better than they could do with all of the federal constraints. Perhaps out of that, if it is tried, we will see the best way to provide health care services. Clearly here, as well as in the private sector and in my judgment, in Medicare, we are going to have to have more elements of competition to control costs, which obviously includes managed care.

Part I:
The Big Picture

World Population and Health

Lincoln C. Chen, M.D., M.P.H.
Takemi Professor of International Health,
Harvard University School of Public Health

WORLD POPULATION CHANGE

This discussion reviews the state of the world's population, its changes, and its implications for world health in the 21st century. My argument is that much of the future is already here in terms of demographic change and the globalization of health. Demography is the study of the size, composition, and distribution of human populations. Although quantitative methods are employed, demography is also centrally concerned with the quality of human populations, such as their health status.

Carbon dioxide emissions on planet earth are fingerprints of human activities. Such fingerprints would show large cities, as in North America, forest fires in the Amazon, and gas fires associated with petroleum production in the Middle East. For the first time in history, our human species has the technological capability to alter irreversibly our geophysical environment. These productive and consumptive patterns have generated remarkable wealth in some countries, although the wealth is concentrated mostly in three major regions—North America, Western Europe, and East Asia. There is a mismatch between wealth and where the world's people reside— overwhelmingly in the poorer regions of China, South Asia, and parts of Africa and Latin America.

In this our 20th century, world population experienced an unprecedented increase, from 1.7 billion in 1900 to more than 6.0 billion by the year 2000. Although forecasts about the future can reflect either "mumbo-jumbo" fortune-telling or statistical probabilities, population projections are neither, but are based upon assumptions about continuations of past trends modified according to assumptions about future changes in birth and death rates. The population projections performed by the United Nations, the World Bank,

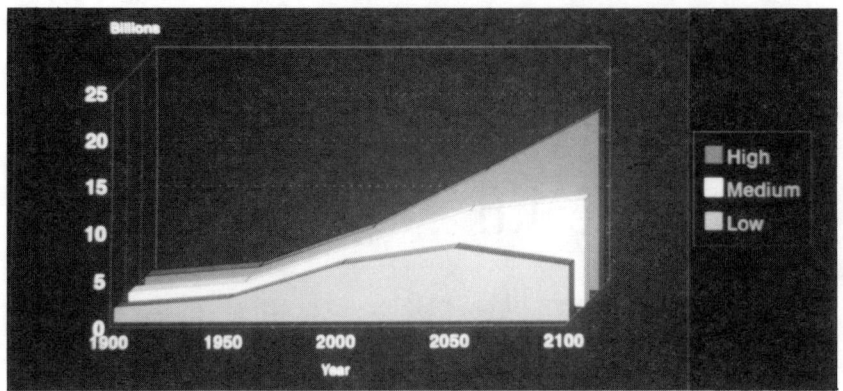

Figure 1. Population projection variants, 1900–2100.

and the U.S. Bureau of the Census all generate, however, remarkably similar demographic futures, as shown in Figure 1. By the end of the 21st century, we can expect a world population anywhere from about 8 billion to nearly 20 billion, depending on high, medium, or low assumptions of slow, medium, or very rapid reductions of human fertility. In these projections, mortality is assumed to continue its long-term secular decline—not an altogether safe assumption given the many human crises we are experiencing around the world. The annual number of new additions to planet earth is now peaking at about 90 million, and although this growth will continue for another decade, it will decline steadily throughout the next century. The peak rate of growth of the world's population was about 2.1 percent per year around 1970, having declined to today's rate of 1.6 percent. The net increment of population increase today is equivalent to adding one new Mexico or three new Canadas each year to the world's total population.

World population growth reflects a remarkable decline in fertility in the developing world, from an average of six children to four children between 1965 and 1985 (Figure 2). An unprecedented pace of decline has been experienced in much of Asia and Latin America. Slower declines were witnessed in South Asia and the Middle East, while little or no decline has been experienced in sub-Saharan Africa, although there is evidence that fertility has finally begun to decline in about four African countries. Throughout North America and Western, Central, and Eastern Europe, fertility has been near or even below replacement. The fertility declines have been associated with a marked increase in the prevalence of contraceptive use, from 9 to 45 percent of all eligible couples in the developing world.

Our demographic future over the next 25 years appears to be already with us. We will definitely have more people in the world, in large measure because of "population momentum," which is built-in growth due to reproduction among children already born. In other words, even if all couples were to achieve replacement fertility immediately, the youthful age structure of the world's population would still generate future growth because the children already born will move through their reproductive years over the next three decades.

We will also have changing population compositions. Some have called the 20th century the "American century," in comparison to the 19th century—the "British century." The 21st century is likely to be the "Asian century" because of the shift in the gravity of people and wealth to East Asia, a process well under way at the close of this century. Whereas Europeans and North Americans constituted 34 percent of the world population in 1900, they will constitute only 14 percent by the year 2100. Asia's share of half the world's population will remain steady, but very large proportionate increases will come from Africa and Latin America. Another aspect of compositional change will be the aging of populations. High-fertility developing populations have a youthful (triangular) age structure in comparison to low-fertility industrialized countries, which have a more elderly (rectangular) age structure. In all countries, because of declining fertility, populations will continue to become more aged.

Finally, world populations will demonstrate changing geographic distributions. Rapid urbanization will translate into megacities throughout the world, especially in developing countries. Urbanization will be accelerated in part by continuing rural–urban migration. International migration also can be expected to accelerate in the 21st century. As described later, international migration may develop into a most important demographic feature of the 21st century.

DEMOGRAPHY IN THE UNITED STATES

High-, medium-, and low-variant projections of the U.S. population into the 21st century result in population sizes between 300 million and 400 million. Here, assumptions about rates of growth are generated mostly by dif-

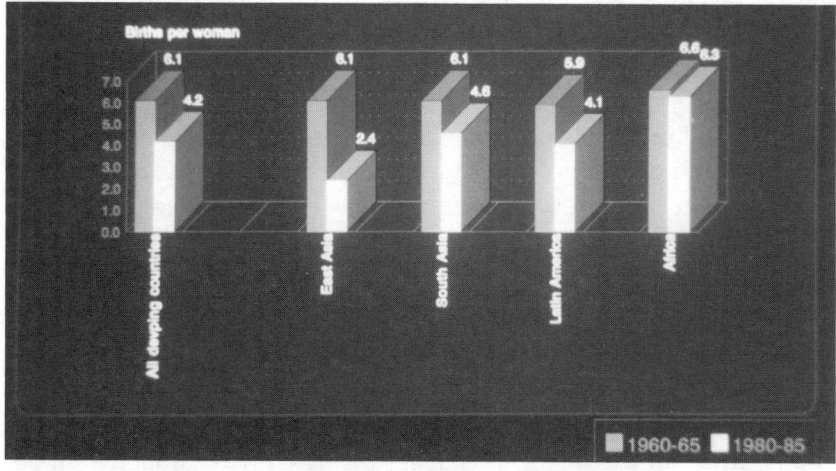

Figure 2. Fertility trends in the developing world, by region.

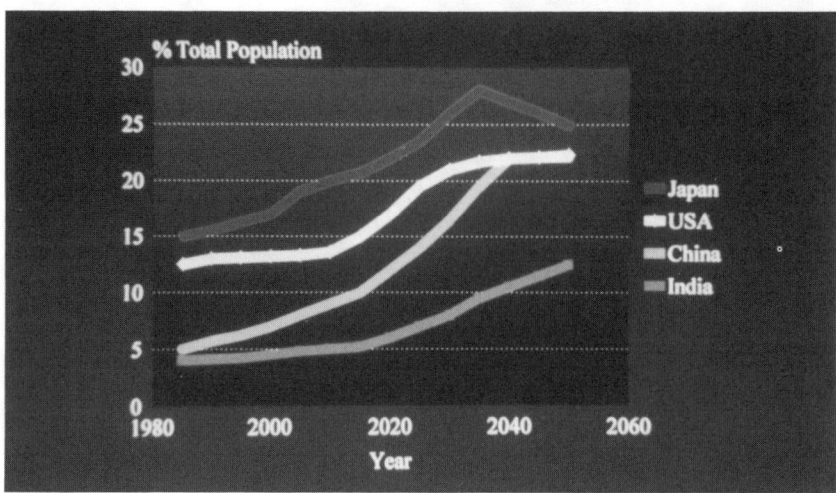

Figure 3. Percentage of population aged 65 and over. SOURCE: U.S. Bureau of the Census.

ferent levels of immigration, not fertility. Rather than population size, it is age structure that will experience the most pronounced change. Figure 3 shows the proportion of the population over age 65 years in the United States, Japan, China, and India. By the year 2020, the United States will have more than 20 percent of its population older than 65 years, higher than today's most aged society, Sweden, at 18 percent. Japan and China, the latter with its one-child family policy, will experience similar aging processes. Among the aged, the proportion of those who are "very old" (over 75 or 80) will also increase. Aging raises concerns not only about the economic and social aspects of care for the elderly, but also about the ratio of elderly dependents to productive adults, whose caring responsibilities will shift increasingly from children to the elderly.

A third demographic feature of America's population change will be migration. Population redistribution away from the Northeast to the Southwest will likely continue. Equally pronounced will be a shift in the ethnic composition of Americans due to immigration. By assuming high rates of immigration, the ethnic composition of Americans from 1950 to 2050 will generate two striking changes. Anglo Americans will be a minority group and the Hispanic population will grow to become almost twice the size of the black population in the second half of the next century. Medium and low levels of immigration will mute some of these changes, but the directions will remain constant.

HEALTH: THE WORLD AND THE UNITED STATES

Demographic change, worldwide and within the United States, will powerfully affect all aspects of the quality of life—the environment, food, the economy, schools, jobs, and health. In world health, this century has witnessed a remarkable revolution. Unprecedented progress has been enjoyed

Table 1. Epidemiologic Transitions

First Generation of Diseases
Common Childhood Infections
Malnutrition
Reproductive Risks

Second Generation of Diseases
Cardiovascular
Oncotic
Degenerative

Third Generation of Diseases
Environmental Threats
　Air, water, chemical
　Ozone depletion, global warming
New/Emerging Infections
　HIV/AIDS, Ebola virus, plague
　Tuberculosis, dengue, cholera
Sociobehavioral Pathologies
　Violence
　Drug abuse
　Mental and psychosocial illness

by all countries, industrialized and developing. In terms of health, our grandparents lived in an entirely different health world from that we enjoy today. It is worth recalling that the infant mortality rate in New York City at the beginning of this century was higher than it is today in Bangladesh. Health advances, however, have been uneven, with many countries, especially the economically poor and politically unstable, being left behind. Infant mortality rates still vary more than a log order between the best and worst countries. The positive picture of progress, therefore, is counterbalanced by concerns about health inequity and health diversity between and within countries.

The heterogeneous health situation is due to the complex epidemiologic and health transitions taking place around the world. Table 1 shows how the epidemiologic pattern of the causes of death changes as life expectancy improves. At low life expectancies, poverty-linked causes of death due to common childhood infections, malnutrition, and risks associated with childbearing predominate among women and children. As life expectancy is extended, affluence-related cardiovascular, oncotic, and degenerative diseases begin to predominate among adults and the elderly.

Some have hypothesized that we are entering an era of a "third wave" of environmental, infectious, and behavioral disease threats that are shared by all societies, rich and poor alike. Even as we have yet to overcome the first generation of environmental health problems due to lack of clean water and sanitation, we are confronting new environmental problems of ozone depletion, global warming, and the safe disposal of toxic waste. Because ecological fragility and population growth threaten the food balance sheet, the age-old question arises of how many people the world can feed. Recently, Joel Cohen at the Rockefeller University reviewed estimates of the earth's capacity to feed its people. He found more than 65 studies over the past centuries that estimated the earth's food production capacity as feeding anywhere from 1 billion to 1,000 billion people, with most estimates ranging from 5 billion to 10 billion. In the review, he underscores that this may be the "wrong question," because the number of people that can be fed depends upon human choices in terms of the quality, style, distribution, and pattern of human consumption. In other words, this issue is not simply ecological but rather moral and within the choice of human agency.

The Institute of Medicine at its 25th anniversary annual meeting underscored concerns over new and emergent infectious diseases, such as HIV/AIDS, Ebola virus, cholera, plague, dengue fever, and the development of multiple resistant organisms. There are, furthermore, a host of sociobehavioral pathologies, such as violence, drug abuse, and mental illness.

This third wave of disease threats may reflect the emergence of health as part of the process of globalization. Acceleration in international trade of goods and services, rapid flows of financial capital through the global private marketplace, the communications and technology revolutions, and migration both temporary and permanent may be generating new transnational health linkages that will create world health interdependence, just like economic interdependence. Transnational connections in health imply that health threats no longer can be contained by national frontiers; most diseases do not require passports to travel. National sovereignty in health is eroding rapidly. Transnationalization, we already recognize, will carry international health processes across national borders. The 1991 cholera outbreak in Peru was probably transmitted by ships moving goods from Asia to Latin America. With the North American Free-Trade Agreement, nursing homes for elderly Americans may find the lower labor and energy costs in Mexico attractive. Communications and transport are compressing distance and time. More than 500 million air travelers cross national frontiers annually, and international migration—family reunification, economic migration, refugees—is likely to grow given the porous movement of investment, production, and consumption. An increasingly literate public will understand the science of health better than at any time in human history, propelling public health to develop its third "public" dimension—public governmental service, population approaches to public health problems, and public agency of its own health. International cooperation, therefore, may become even more essential for national health.

CONCLUSION

I close by underscoring three dimensions of world health and population in the 21st century. First, health is both part of the problem and part of the solution to world population change. The unprecedented growth of human populations in the 20th century was due demographically to the very rapid decline of mortality in relation to fertility, especially in developing countries. Yet, as recently articulated at the Cairo population conference, the provision of basic reproductive and child health services and women's education and empowerment constitute the most effective approaches to population stabilization. Universal access to affordable, high-quality reproductive and child health services—contraception, control of sexually transmitted diseases including HIV/AIDS, maternal and child health—offers the most effective and humane approach to attaining good health, enabling couples to achieve smaller desired family sizes and accelerating the world's demographic and health transitions to stability and quality.

Second, America's health in the 21st century must wrestle successfully with equity between the young and the aged and among social and ethnic groups. The aging of America's population will have enormous social implications for family structure and caretaking of the elderly, economic implications for health care costs, and will also affect our body politic. Imagine a voting population one third of whom are older than 65. President Clinton and Speaker Gingrich have been wresting with this phenomenon in recent debates over the future of Medicare and Social Security. The politics of health care access will also be accentuated by immigration and the changing ethnic composition of America's population, as the Anglo majority declines and Hispanic and other minority groups increase. How the poor and disadvantaged in the United States gain access to high-quality services, through programs such as Medicaid, is very much a part of this debate about national priorities and responsibilities.

Finally, America's health in the 21st century is already inextricably linked to world population and health. We are becoming part of a "global health village" because of health interdependence and the transnationalization of disease. Most health problems are commonly shared, and many health risks clearly have transnational properties. The imperative for international cooperation will intensify.

For the Institute of Medicine in its next 25 years, three questions are central: (1) How can America's powerful scientific capacity be shared with the rest of the world? (2) What can we learn from successes and failures from around the world? (3) How much isolated progress can we enjoy without parallel progress elsewhere? Health in America will not prosper if it cuts itself off from the wealth of scientific opportunities elsewhere, fails to learn from the successes and failures of others, and strives in isolation to advance health while others flounder. These are not purely academic questions; they are also practical and moral. I have no doubt that at its 50th anniversary celebration, the Institute will be able to report that its "2020 vision" has not only reacted to but led the world and the American public in addressing these questions for the sake of our common health.

DISCUSSION[*]

PARTICIPANT: How do global income disparities affect health?

DR. CHEN: We are witnessing global economic transformations in all countries. I think we must recognize that the private market, for all of its efficiencies, is not and can never be equitable. There needs to be an ethical basis and a social set of relationships underlying the equitable and efficient functioning of economies that are based on capitalism. We have not found a solution to that in the United States, as we search for our future, and the world is pursuing the same approach.

Now, there are conflicting data about what this means in terms of health conditions. On the one hand, there is a convergence in the rates of improvement of life expectancy across countries. Yet these data show a divergence in terms of capacity to command income between countries. So you have a diverging income stream and a converging life expectancy stream.

The best theory to explain this is "disguised morbidity." Because of knowledge, we are more able to avert death, but we nevertheless carry heavier burdens of disease among those who are disadvantaged.

Second is the capacity of people with knowledge to protect themselves against mortality. I think that this is one of the real contributions of medical science in the 20th century. We can protect ourselves against mortality, but we cannot necessarily protect ourselves against sickness.

PARTICIPANT: What about cross-cultural values in shaping health in different cultures?

DR. CHEN: I certainly agree that the multiculturalism in the United States and internationally will pose many ethical dilemmas. Our technological capacity obviously is running far ahead of our moral and ethical consensus about how to deal with these questions, both domestically and internationally.

I am going to a workshop in February in Japan that will deal with Asian values and cultures with respect to health and population change in the next century. Some Japanese and Chinese, for example, feel that they have unique ways of viewing these issues.

I will mention two other dimensions very quickly. One is that there is a growing universalism in the recognition that good health is a human right, not just a privilege. This is something that I know is debated in the United States. We tend in American culture to view civil and political rights as human rights, but do not view socioeconomic matters, including health care, in terms of rights. Other cultures actually have more balanced views about civil or political and social or economic matters.

---

[*]Because of technical difficulties, the participants' questions had to be reconstructed; therefore, they are not verbatim.

In your question, you were perhaps referring to the second dimension, which involves Sam Huntington's theory about the "clash of civilization and culture." In his theory, the future of world politics will not be a north–south or east–west clash, but a clash between large cultural civilizations.

PARTICIPANT: What are the differences between Asian and Western cultures in terms of care of the aged?

DR. CHEN: One difference is that, for example, Singapore, Japan, and China—historically and at the present—have made policies that they will go into the future with a family-based rather than institutionally based systems of nurturing and caring for the elderly. So they are not setting up institutionalized systems. They are counting on three-generational family residential capacity to manage the aged.

Now, this is fine if you are a man, but not if you are a woman. I gave a speech on this topic in Tokyo to an audience of Japanese men about 5 months ago. Actually, they took it quite seriously, because women are entering the work force in very large numbers and want to participate in the political, economic, and social life of the society. So the current policy, for example, in Asian society is to have a family-based system of care backed up by communities and institutionalized care. Yet, I am uncertain that Asian women, who would be required to manage the burden of care of the elderly, would be willing to accept such sacrifices. There has been very little work that I know of about the cost of technological intervention in the elderly age groups in these societies.

I might add that Asians, in the cultural area related to aging, feel very strongly that they have veneration and respect for the aged that has been lost in Western culture. However, in historical England, the same type of respect and veneration in the family toward the aged can be found, but Western culture has somehow lost these values. The Asians at the Japan conference, for example, were very sensitive to the fact that while something like this was lost in Western culture, it is not something that Asian cultures want to lose. I am a little skeptical, however, as to whether they will be successful with a family-based strategy.

PARTICIPANT: Could you comment on injury-related health problems?

DR. CHEN: We had a workshop at Harvard recently on the global burden of violence, and world homicide and suicide statistics were presented. Homicide rates are invariably higher among men than women around the world, but the higher female suicide rate in China was a puzzle. Actually, some of the Chinese who come from Beijing spoke to this question in terms of the ways in which young Chinese women are trapped into looking at their future without social choice and opportunity.

Information and Communications

Don E. Detmer, M.D.
Senior Vice President, Louis Nurancy Professor of Health Sciences Policy, and Professor of Surgery, University of Virginia

I greatly appreciate this opportunity to share my sense of what lies ahead, including a few ways in which the Institute of Medicine may help shape the information age. My discussion is about looking forward to 2020, which is a presumptuous task. What you might get instead is a "my-opic" view of 2020. At any rate, we can say that there is something quite exciting coming.

Lewis Thomas would remind us that our biology is based upon a carbon structure. Now we are entering an era in which we will fuse carbon-based intelligence with capabilities that are silicon based. We are crossing the threshold into a new chapter of human existence. What follows are my speculations on this inspiring fusion.

Vannevar Bush stated some years ago that the world had arrived at an age of cheap, complex devices and great reliability, and something was bound to come of it. The status was no longer quo. As Bela Bartok put it, "What is new and significant must always be connected with old roots, the truly vital roots that are chosen with great care from the ones that merely survive."

"Bite by bite and bit by bit," as Robert Kahn said, "eventually adds up to quite a bit." Today, computers can store and retrieve a thousand-trillion bits of data, and the expectation is that there will be performance at the 1 million-trillion level by the end of this decade. Petabytes and etabytes are quite a bit.

Byte by Byte/Bit by Bit

byte =	8 bits
kilobyte (10^3) =	1,000 bytes
megabyte (10^6) =	1 million bytes
gigabyte (10^9) =	1 billion bytes
terabyte (10^{12}) =	1 trillion bytes
petabyte (10^{15}) =	1,000-trillion bytes
etabyte (10^{18}) =	1 million-trillion bytes

. . . quite a bit

Figure 1. Computer power continues to increase with a quadrupling of power every four years. Petabyte levels have been achieved, and etabyte levels are expected by the end of the decade.

The big picture, as I see it, is in terms of both information and communication, comprising the medium—the telecommunications infrastructure, as well as the message—managed care capitation, biomedical research innovation, public accountability, and the topic that I am principally addressing, how all of this fits together.

The capability will become powerful enough for some of these distinctions to blur at the margin. So I want to mention a few high points related to the information infrastructure itself, the medium, and then focus on a few points related to the messages before offering some philosophical reflections on computer ethics.

The major message for health professionals using a telematic medium will relate to managed care (capitation), biomedical research and innovation, public accountability, and the technology itself, including the computer-based patient record as envisioned in the 1991 IOM report, *The Computer-Based Patient Record: An Essential Technology for Health Care.*

The infrastructure is changing rapidly and, at the same time, not as quickly as people perhaps would like to think. We are seeing more personal computers with mainframe power, scalable parallel high-performance computing systems capable of monitoring and maintaining a superpower's nuclear force without future actual testing, and the emergence of a global neural network connecting the U.S. national information infrastructure to the G7 (Group of Seven) global information infrastructure.

More than its share of leadership for the global information infrastructure is coming from the United States, stimulated by the development of the U.S. High-Performance Computing and Communications Program (HCCP) enacted in 1991. This act, championed in 1989 by Vice President Gore, then Senator Gore, has given focus and integration to the computing initiative of 10 separate governmental agencies. As most of you know, IOM Council member Donald Lindberg was the founding director of the HCCP and as head of the National Coordinating Office gave health applications the visibility and the attention that they certainly deserve.

Four Transformational Forces

1. Managed care or capitation systems that place providers at risk.
2. Biomedical research and innovation.
3. Public accountability.
4. Telemedicine and the computer-based patient record.

Figure 2. These four major forces will transform health care over the next score of years. If we manage these forces and their interactions, substantial increases in the quality of life can result.

If we look at the global situation today, we see a patchwork of incompatible communication networks marked by high-cost, low-quality services and very limited interoperability between systems. However, that infrastructure is changing with the emergence of broadband transmission.

Broadband transmission—the cable, wire, and wireless transmission of multiple frequencies of more than 2 million bits per second—is the cement that will bond communications and computing. The major issues about the technology itself are still not settled. For instance, will the communication mode be in bits routed through asynchronous transfer mode (ATM) switches or in waves from variable-wavelength lasers. According to some experts, if the bit mode prevails, fixed-length, 53-byte cells will be routed by way of ATM switches, which will combine the best of circuit and packet switching with efficient bandwidth guarantee if the standards are in place. The estimated time of delivery is 2000 or 2010.

Alternatively, optical ether transmission in wavelengths tuned to varying wavelengths by lasers would travel instantaneously by way of electronic switching equipment, with the bit technology totally eliminated. The estimated time of delivery for this is 2010.

When will multimedia really be here? Without demand, truly global broadband networks could be decades away; it depends upon consumer demand for services. Short of massive government subsidies, American-led entertainment and recreation are likely to lead to its development, with business applications and health care following in succeeding waves. The truly ubiquitous broadband networks across the United States could even take 30 years.

So, despite the challenge of creating a global network, we are seeing principles for a global information infrastructure emerge, as well as global organizations such as the International Association for the Advancement of Health Information Technology, which was initiated in Geneva this fall. Overcoming the political, cultural, and financial barriers is part of the challenge, but leadership remains the most crucial aspect. We need more leaders such as Don Lindberg, Jan H. Van Bemmel in the Netherlands, and Marion Ball, who recently completed an outstanding term as president of the International Medical Informatics Association.

The principles include interoperable systems and applications, privacy and data security, protection for intellectual property rights, universal access to networks, and the encouragement of research and development. Despite what is yet to be done, much is happening. For example, the Internet is expanding rapidly. In March of this year, the High-Performance Computing Center reported that there were 27,000 networks, representing a 350 percent increase in the past 12 months alone.

Forty countries are now connected by way of the National Science Foundation's International Connection Management Program. Estimates vary widely from 10 million to 24 million people who are now currently connecting with the West, but no one denies that the number continues to grow.

Telematics is a recent term used to reflect the impact of the coming broadband capability. Whereas the video component today is 10–15 frames per second, camera and voice require 90 megabits a second for a normal appearance.

One way to define health telematics is that it encompasses both telemedicine and medical informatics or health informatics. Telemedicine and its evaluation are topics of a current IOM study chaired by John Ball.

In addition to teleconferencing for educational purposes, telemedicine also helps transmit computer-based multimedia patient records for consultation over large distances. Early experience suggests that substantial dollar savings are realized by better triage decisions regarding transfer of patients to higher level care facilities.

Virtual surgery is also in the wings. In Finland a few weeks ago, a surgeon from France reported at a meeting I attended that a pig in San Francisco had its gallbladder removed by a surgical team working in Texas and Oregon. I have been unable to confirm that story, but whether it is true or not, virtual surgery is on its way. Beyond treating acute appendectomies on submarines, it is not quite clear what the implications of this might be.

For those of us who live in states with substantial rural populations, telemedicine offers a real potential to get health care services to isolated communities and into people's homes generally, particularly for the elderly.

Medical informatics is the use of computers and information sciences to facilitate clinical practice, medical research, and teaching. Examples of applications include knowledge bases, calculations, data retrieval and display, decision support and evaluation, synthesis, and analysis.

The growth in the knowledge base of medicine alone is prodigious. Reports by John M. Williamson and others have shown that only about 10 percent of primary care physicians keep up with the medical literature. I cannot find similar studies on other types of physicians, but it is clear that only with information and communication technology can both generalists and specialists, as well as all members of the health team, hope to stay current with the exponential growth of the knowledge base.

Further, it is increasingly impossible to keep up with the paper and the digital information glut, especially as we live in this messy "metastate" with both paper and digital systems. Our drive to become fully digital is now increasing in fervor as we struggle to reduce the deluge of overlapping and duplicated information.

Herbert Simon, the Nobel Laureate, captured this nicely:

> What information consumes is rather obvious. It consumes the attention of its recipients. Hence, a wealth of information creates a poverty of attention and a need to allocate that attention efficiently among the overabundance of information sources that might consume it.

Where do we place our attention so that these information and communication technologies can best improve human health? This is going to be the balancing act for the next 25 years.

Clearly, we need to relate telematic technology to the messages of our environment: managed care, capitated care, biomedical research and innovation, as well as public accountability. Yet, more specifically, how do we do this—particularly since the complexity of our world is such that we will not control it?

For example, all of us are managing, or trying to manage, managed care, but it is obvious that we are dealing with forces we do not ultimately control. As Woody Allen says, "The lion and the lamb will lie down together, but the lamb won't get much sleep."

So I want also to discuss briefly the shadow side of this. Will the medical marketplace focus on price or value, on profit, or on access to needed services? Will telematics be built to serve the economically advantaged part of the world or everyone? Will its software programs achieve the stated goals or simply represent "vaporware," that is, software whose performance does not come close to its advertised value? Will our standards of accountability in medicine be our clinical impressions, or will they be based on hard evidence? Finally, will our patterns of interaction reflect continued paternalism or maternalism, or will we use these information communication technologies to develop real partnerships with our patients and society?

I would posit that between now and 2020 we have a grand challenge with two component parts. The first is to mature the health evaluation sciences: biomedical science, biotechnical sciences, ethics, biostatistics, clinical epidemiology, health services research, and telematics. The second part of this grand challenge is to integrate the health evaluation sciences with the evolving biological and the psychosocial science base.

I think we can attend to that challenge only by capitalizing upon information and communication technology. This will require us to develop and manage knowledge bases, to manage value, to ensure accountability through evidence-based medicine, to integrate patients' desires into our practices and public reports through the development of computer-based records and computer-based record systems, and finally, to formulate methods and policies to critique our goals and values.

Any sensible development of knowledge bases must have robust vocabulary servers with a unified medical language. A multinational effort mounted

Medicine in 2020: Grand Challenges

PART I

Advance the health evaluation sciences, including:

- Biomedical/biotechnical ethics
- Biostatistics
- Clinical epidemiology
- Health services research
- Telematics (telemedicine and health informatics)

PART 2

Integrate the health evaluation sciences and the biopsychosocial science base

Figure 3. Over the past 70 years, the basic sciences of medicine have developed. Medicine's success in the next 70 years will come from developing the health evaluation sciences and integrating them with the continually developing biological and psychosocial science bases.

at this time would avoid a great deal of effort to pull divergent approaches together in the future. A major effort begun now could essentially be complete in five years, including setting up a process to maintain it into the future. The IOM could play a major role in this important global development.

Knowledge bases at various levels and for a variety of purposes already exist. These include some, however, totally new and unique data sets. For example, with emerging on-line communication, observational data shared among patients themselves are developing real credibility, particularly among the patients. What are we as professionals and the patients as patients to make of these databases? Can we examine these data and reshape them into knowledge bases that specialty societies take on as part of their mission? Clearly, the task of maintaining knowledge bases is likely to be even more daunting than their initial development.

As the excellent IOM definition of quality of care illustrates,[1] if we are to manage to value instead of price, the clinical knowledge bases will never be completely static. Current professional knowledge is certain to be a swiftly moving target because of the contributions of biomedical research and innovation. Maintaining and evaluating our knowledge bases will require constant attention. Specialty organizations again should serve society by performing this admirable function. In addition, as the World Wide Web becomes filled with all manner of health-related data, whose role will it be to sort out the signal-to-noise ratio?

---

[1]"Quality of care is the degree to which health services for individuals and populations increase the likelihood of desired health outcomes and are consistent with current professional knowledge." From *Medicare: A Strategy for Quality Assurance,* 1990.

With respect to the delivery of health care as we know it, greater public accountability and involvement are needed in the care of individual patients, as well as in reporting the results of the care of groups of patients.

Report cards on organizational performance, such as those being issued regularly in Cleveland, come to mind. Currently, these require a great deal of effort involving paper systems. The computer-based approaches that follow this will be more efficient and complete.

Only with robust information systems can we get more science into our procedure-based clinical work. Evidence-based medicine of this type has had three principal reports: Archie Cochran estimated in 1976 that less than 10 percent of medicine is currently grounded solidly in reliable data; the Office of Technology Assessment (OTA) in 1983 estimated 10–20 percent; and NIH in 1990 estimated 21 percent. Whatever the exact situation, the situation is improving but still has a way to go.

The greatest impact of information technology on practice should come from the momentum developing behind evidence-based practice. This will involve automation of a number of care processes, and only time will tell how quickly these will develop. With the proper set of incentives and data systems, the changes could be surprising.

The possibilities include technology assessment, management of diffusion, clinical guidelines, and the debridement of useless procedures from studying what comes out of computer-based records and outcome analyses. The proposed IOM study on assuring clinical proficiency is very timely in this regard.

The Institute's 1991 report on computer-based patient records laid out a vision that included a primary recommendation that such records become the health records of choice. That report, with its 12 attributes for computer-based records and their systems, has generated a great deal of momentum, and the report of another IOM committee chaired by Roger J. Bulger, *Health Data in the Information Age: Use, Disclosure, and Privacy,* published in 1994 maintains the momentum. While progress continues, it is likely to take until 2020 for widely available computer-based records to meet all of the criteria developed by the IOM committee.

One of the recipients of the 1995 Nicholas Davies Award for Excellence awarded by the Computer-Based Patient Record Institute was the University of Utah group at the Latter-Day Saints (LDS) Hospital in Salt Lake City. Reed Gardner spoke directly to what his group has learned from two decades of effort in building computer-based patient record systems. These were his messages: First, improve the process and remind yourself constantly that the patient is central—not the doctor, not the nurse, not the administrator, but the patient. Build systems to prevent problems, not to detect problems that you then have to figure out how to prevent later. Focus on the system, not on the individual. The Utah group also concluded that variation in clinical practices is endemic and that quality can be improved constantly.

Evidence would suggest that impressive gains can be achieved with fairly simple systems that let you practice good medicine but remind and support you until the task gets done. At the individual clinician level, this will also involve development of computer-based decision support systems that integrate the individual patient's utilities into the patient's care plan. This is real accountability at the point of delivery of care.

William Knaus and I have recently discussed how this might be done to improve the care and decision support for dying patients, a need highlighted by the Robert Wood Johnson Foundation-funded study that he codirected, which was recently reported in the *Journal of the American Medical Association*.[2]

Now, let's return to the big picture, keeping in mind Bela Bartok's comment.

I have hinted about a "pro-fusion" of applications and considerations; alas, there will be some "con-fusion" as well. What follows are my speculations. First, I think that society will reformulate around new functional aggregates of geographically dispersed individuals. I am talking about virtual reality becoming actual reality for these subgroups through use of the World Wide Web.

A new, powerful form of ethnicity will develop around telematic-based work and function. This will have its geographic movement component as well. Keep in mind that each day, 1 million people cross national boundaries through air travel. We are just beginning to see elements of this, but I do not think it is yet truly developed. This development will eventually challenge some aspects of nationhood and of cultural and religious heritage.

Cultural concepts will begin to be redefined: for example, democracy, health, and autonomy. A new reality and a redefinition of democracy may emerge that will be a more true democracy rather than a representative one since communication technologies allow for more direct governance, an impossibility without this technology.

Health itself may become redefined to include not only one's unassisted and unaccompanied physical and mental well-being but also a well-being that is intrinsically supported by a number of artificial computer-based capabilities. The inanimate becomes incorporated as essentially being or becoming partly animate. For example, artificial hips, pacemakers, and even imbedded intelligence may emerge. Is a person "healthy" even if he or she is healthy only with a host of new body parts?

Then there is autonomy. When a decision support system mandates that a doctor take action in a tense clinical situation, yet does not reveal how the system arrived at its directive, who is really taking the action, the computer or the physician? Where is the autonomy and where is the responsibility?

How much of our sense of the word "knowledge" will become based not upon the reliability of something's appearance, which we have generally held as a criterion of knowledge, but instead upon meeting a more rigorous test of not only reliability but validity?

---

[2]The SUPPORT Principal Investigators. A controlled trial to improve care for seriously ill hospitalized patients. *J. Am. Med. Assoc.* 274(20):1591–8, 1995.

Consider the blurring of the margin of animate and inanimate, of perception and reality. If you are a surgeon operating in a virtual telesurgical environment, what is real—the image you are working with on the screen and what you assume is happening on the other end, or what is actually happening on the other end? Is perception reality, and when may it not be? What is real? Is the idea itself the reality, or does it become real only when it is physically manifest?

According to M. J. van den Hoven at Erasmus University in Rotterdam, some of these issues are old wine in new bottles, but others appear to be unique to computing technology. In a real environment, ideas and physical reality are few. We can concretize our abstractions. In the near future we will have new abstractions to concretize, or we must figure out how to live abstractly in an abstract world. Lawrence Weed has said that we think in the abstract but we live in the concrete. Tomorrow this may not be true.

Unfortunately for us, as Edmund D. Pellegrino and David C. Thomasman said, "Medicine does not have the tools within its methodology to critique its goals and values." We must build them as telemedicine and telematics force us to create policy to meet new realities. Our information technology policies need to reinforce the most vital roots of our moral philosophy. It is now self-evident that the marriage of information and communication technologies to health will force the IOM, government, business, journalists, and citizens to consider issues and policies that previously have been ignored, as well as to engage newly emerging realities. An example is the currently proposed legislation for confidentiality, accuracy, and the integrity of personal health records, the Bennett–Leahy bill, S. 1360. Even more is at stake, however. Similarly, telemedicine is forcing us to look at our legal structures and political boundaries.

We need to develop new methods to critique our technology, including information and communications technology. I prefer that we call this effort biotechnical ethics, in contradistinction to biomedical ethics. The focus of biotechnical ethics is upon the technology of the information communications revolution as well as other technology used by health professionals.

For example, let's seek ways to shape the design of future software, as well as our terminology, in order to achieve more moral ends than are likely to be achieved if we simply leave events to occur randomly.

With respect to terminology, consider some of the current language that is used in computing applications. Security systems are constructed with what we call "fire walls" to enhance security. Might not "privacy screen" send a clearly different cultural message and expectation? Would "house keys" rather than "password" send a different message?

Beyond these fascinating discussions and interesting word games, we must make certain that we do not pass up the opportunity to shape the future of this powerful technology, since failure to do so could be at the cost of our collective soul. While the tasks are major, the times complex, and the pace swift, the opportunities are tremendous. As John Shaw Billings stated in 1913, "there is really nothing difficult if you only begin—some people contemplate a task until it looms so big, it seems impossible, but I just begin and it gets done somehow. There would be no coral islands if the first bug [had] sat down and began to wonder how the job was to be done."

At its base, our work is honorable. Perhaps over the next 25 years, we will reinforce the best of our values and transmit them to the larger society. I hope so.

DISCUSSION[3]

DR. SHINE: How can we get clinicians to focus upon public account-ability and performance measurements?

DR. DETMER: That is a very good point because these methods are not that new. I have two stories in response. One involves the thoracic surgeons in Pennsylvania who were so disgusted two years ago by the absence of se-verity adjustment in reports issued by the state, that they got into the act as a society. In fact, they built a much more rigorous severity adjustment tech-nology, presented it at the American surgical meeting last year, and essen-tially have owned this issue. In fact, now it is their issue, as well as being a public issue.

Similarly, it appears that in Cleveland, enough of the entire commu-nity—both providers and buyers—is paying attention to this that cost and quality are moving in the direction you would hope to see. By the way, I think Reed Gardner and the people at the LDS hospital are convinced in-creasingly that the very best quality care happens also to be the least expen-sive care.

Clearly, physicians, in order to change their behavior, must feel that data are timely and reliable, for starters. However, incentives should support this behavior as well.

I think that we are turning the corner on this, but the grand challenge is how to move from where we have been into this new era. This is doable and we are starting to see it in places, but how quickly it will spread, I don't know.

PARTICIPANT: How will confidentiality be maintained in the future?

DR. DETMER: The previous speaker mentions our living in a global village. I grew up in a village in the middle of Kansas, where everybody knew everything about everybody else. There was no such thing as privacy; you could hope for confidentiality.

I would argue that today, if people think they have privacy in the United States, they simply have not been of enough interest to the press. In many respects, actual privacy has probably always been somewhat elusive. On the other hand, I think we can greatly improve confidentiality. The basic point is

---

[3]Because of technical difficulties, the participants' questions had to be recon-structed; therefore, they are not verbatim.

that people need to be on notice that when they are dealing with important data about someone's personal life, they have to treat it that way.

The main thing we lack at present is a national law that puts people on notice in a uniform way across the country. I think we can go a long way toward this, but some of it is where we as a society choose to come down on essentially respecting one another more.

PARTICIPANT: How will advances in information technology affect continuing medical education in the future?

DR. DETMER: Being from the University of Virginia, I will put this in a Jeffersonian context. Thomas Jefferson laid out for our country, and in many respects for the world, a system of education that was built of primary, secondary, and higher education. The assumption in that era was that once you got your basic education and went out into the world, you were prepared. You had gotten the skills and the means to go forward.

The fact of the matter is, as you saw by the growth in the knowledge base, we are increasingly reinventing ourselves on-line as we work. This country has not yet developed a strategy not just for continuing medical education but for continuing education in general. Increasingly, it becomes something our society ought to do.

I would favor a form of worker benefit that gives people so many credits for the amount of work that they do, along with the necessary educational time to get away and study.

Your question, I think, is very profound and extraordinarily relevant for a free society. As Madison also said, education literally is crucial to a free society. So it is may be a way of catching it on the wing, the way it is done now, but also developing systems that can actually help us do this. Now, clearly, information technology is going to help. A lot of people may have different views on that, but it is a key question, in my opinion.

PARTICIPANT: How will physicians use computers as "real-time" sources of information, not just in recordkeeping?

DR. DETMER: Gorman et al.[4] published an interesting study last year showing that when doctors see patients, it is very common for questions to arise at that time, but extraordinarily uncommon for the doctors to look up the needed information. Now, if doctors have someone to do this for them or they have a decision support system, the information could be obtained while they are seeing the patient. Relevant data are found more than half of the time and, 80 percent of the time, will change what the doctor then does for the patient. We have a tremendous opportunity to leverage care if we can figure out how to access point-specific data and point-specific issues at the necessary time. However, the system has got to be lean, clean, and pretty mean.

---

[4]Gorman, P. N., Ash, J., Wykoff, L. Can primary care physicians' questions be answered by the medical literature? *Bull. Med. Lib. Assoc.* 82(2):140–60, 1994.

I do not know the nursing database, but doctors typically like to get information, first and foremost, from a direct encounter with another doctor or another person. Telephones are next in line, and computers follow in terms of preference. So the broadband capability coupled with decision support systems may be helpful. It would appear that we have tremendous leverage in this.

By the way, textbooks, which are often a source to which people go, are usually sufficiently dated by the time they are published that they do not reflect what is in Medline and the databases of the National Library of Medicine, and, hence, most current expert practice.

PARTICIPANT: Where will the support come from for training the health care work force on all of the advances in information systems?

DR. DETMER: I would hope that some of the private foundations would see this as an absolutely essential investment opportunity and need, because it is part of the grand challenge. We must invest in greater education; the National Library of Medicine is doing what it can, but I think we really need to get more help on this from the foundation side as well.

PARTICIPANT: How will physicians develop trust in these new systems?

DR. DETMER: It turns out that if you have something that works and you do it right the first time, ultimately, that is best. Although that is a rather bold statement, I think there are growing examples of where it appears to play out in clinical medicine.

PARTICIPANT: Right now there seems to be a lot of marketing of services rather than sharing accurate information about outcomes. How will consumers sift through and make sense of all this information?

DR. DETMER: I think we are going to see a number of changes occur as both the media and the message move toward reality. With respect to the basic language—the terms, vocabulary, syntax, and some of those things— we are not where we could be today, whether we communicate through the Internet or any other way; and that is really what I was calling for.

PARTICIPANT: With all the focus on technology, how will issues like equity of access to technology be addressed?

DR. DETMER: This has been a worry—it is partly why I am interested in biotechnical ethics. The fact of the matter is that the technology can be sufficiently captivating and dazzling, but it can become confused as an end rather than the means to desirable ends.

If the computer interfaces are shaped properly, we will actually free up some professional time so we can deliver new treatments and learning with a human face. There is also a risk because humans are very curious and easily fascinated; we can get caught up by the glitter and think it's the goal. This whole notion of what is a means and what is an end needs to be kept in clear focus.

PARTICIPANT: What impact will all of this additional information have on health care costs?

DR. DETMER: This is anybody's guess right now, because it is a very democratic Net out there. Anybody can throw out anything into the Internet. As I said, it can be noise or it can be signals. The problem is, of course, what all of that does. If it generates more demand for services that are off the wall and trivial, it does not help matters.

That is why I think we are calling for a new challenge to the health professional to try at least to figure out a way to address this new wave we are moving into, because with some of these devices I can see as much opportunity for mayhem as for improvement.

PARTICIPANT: It is unclear how education and research will be supported in a market-driven system of health care. How will all these advances in information technology be supported in medicine?

DR. DETMER: Looking at the current scope and scale of information technology in health care, we are underinvested within our industry by a factor of three compared to airlines, banking, finance, and such. Candidly, it is hard for me to say exactly what the effect of that will be, but it obviously could be profound. Health care is an information-driven line of work.

So as we scale up in information technology, there is absolutely no question in my mind that it will have a profound impact on how we organize ourselves. However, I really have not taken sufficient time to look at more of its organizational implications.

Global Health

Richard G. A. Feachem, CBE, PhD, DSc(Med)
*Dean-Emeritus, London School of Hygiene and Tropical
Medicine, and Senior Advisor, Human Development
Department, The World Bank*

I would like to pick up the story of global health and global population change, begun by Lincoln Chen. As he stated, the life expectancy picture that we have seen in recent decades has been quite remarkable. One way to think about where we are in human history is that in the beginning, there was high mortality and high fertility; and in the end, there will be low mortality and low fertility.

The interesting thing is that the beginning was quite recent, probably about 1800, and the end may be about 2050. The transition through which humankind is passing, from high mortality and fertility to low mortality and fertility, is occurring in this brief time span, and it is occurring more rapidly in the decade in which we are living and working than at any other time.

The changes in life expectancy in the world have been rapid in recent decades, and they have affected not only the more prosperous regions in the world—for example, the established market economies—but also the poorest regions of the world, where life expectancy has also increased greatly.

The region of the world with the poorest economic performance in recent decades, sub-Saharan Africa, has nonetheless gained greatly in life expectancy over the last four decades. Another way of putting this is that the life expectancy of humankind overall has increased more in the last 40 years than in the previous 4,000 years.

This rapid change in life expectancy and mortality is going on at the same time as a rapid change in fertility. Again, we are privileged and challenged to live in a period of human history when the rates are exceptional.

As recently as 1955, there were two worlds. There was a high-fertility world and a low-fertility world. Today, we no longer have two worlds. We have the gradation that Dr. Chen referred to. The low-fertility world has stayed low fertility, but the high-fertility world has broken up, with some regions, such as China, coming very close to the low-fertility world of the more wealthy countries and all regions except one showing remarkable declines in fertility.

The one exception is sub-Saharan Africa, but fertility rates in Africa have now begun to decline.

Changes in life expectancy are obviously brought about by changes in particular age-specific, cause-specific mortality rates. As you are aware, the greatest engine of life expectancy increase is infant and child mortality decrease. In 1960 there were many countries in which more than 17.5 percent of the children born alive never reached the age of five. Today there are fewer and the very highest infant and child mortality patterns have become primarily an African phenomenon.

Clearly, the challenges remain great, the inequities remain very large, and there is absolutely no room for complacency. As we think of those challenges, of the dramatic progress made in the last four decades, and of the progress that still needs to be made in the decades leading up to 2020, we first have to remind ourselves that the major influences on population health are not health professionals and do not lie within the mandate of the health sector. This is a sobering thought.

Those of us who work in health and the health sector need to bear in mind that the levers in our hands are not the most powerful ones. The most powerful levers are four. First, reduction of poverty at the national level, the community level, the family level, and the individual level has been and will continue to be the greatest engine for health improvement and for fertility reduction.

Second, improvements in education will go hand-in-hand with that, particularly for girls' education. Despite a decade of giving lip service to the education of women, disparities around the world between the education of boys and girls remain wide.

Third, fertility decline will make an important contribution to the improvement of health, and the improvement of health will make an important contribution to fertility decline.

Last, there is the environment—not those aspects of the environment that the green movements in this country and elsewhere like to emphasize. Not air pollution, global warming, the ozone layer, or biodiversity. In health terms, these are trivial. I refer here to the domestic environment—the water supply, sanitation, indoor air pollution, and drainage in and around homes—because these things have a large influence on the health of the majority of the world's population. As they improve, so will health improve.

Now, curiously enough, when a team of people gathered in 1991 and 1992 at the World Bank to write what became the 1993 World Development Report entitled *Investment in Health,* they discovered to their amazement that there was no map of mortality for the world. When I talk about a map of mortality for the world, I mean a disciplined statement of cause of death by region, by gender, and by age. If you surveyed experts in various parts of the World Health Organization about particular causes of death, and you asked people dealing with diarrhea how many diarrhea deaths there were; people concerned with pneumonia how many pneumonia deaths there were; and people working on malaria how many malaria deaths there were, and so on, the figures would add up to about two times the total number of deaths in the world. For this reason, we set about to create a disciplined map of mortality in the world, which allowed only the true number of total deaths.

The outcome for the developing countries in 1990—that is, all countries minus the OECD, was 39 million deaths. The unfinished agenda of childhood infections is represented by the second most common cause (pneumonia) and the fifth most common cause (diarrhea). However, the new agenda of noncommunicable diseases and injury is not something to worry about in a few years' time. It was with us in 1990, with cardiovascular disease in first place, cancers in third place, and injuries in fourth place.

As many of you are aware, the use of mortality as a way to measure the health status of the population is unsatisfactory because it downplays causes of ill health that may be very important but that lead to low mortality. One of the most common causes of morbidity in the developing world—skin infections, which cause almost no mortality—does not show up on mortality statistics but nonetheless is important. Another example is mental illness.

Therefore, a new measure was constructed—the DALY, the disability-adjusted life year, which combines morbidity with mortality and enables a burden of disease to be defined that can be used across age groups, across genders, and across causes. All causes of illness have been divided into three groups: Group 1 represents communicable and reproductive causes; Group 2 represents noncommunicable diseases; and Group 3 is injuries.

In the younger age groups in which the burden of disease is high, the unfinished agenda—Group 1 causes—dominate the picture. In middle age, the Group 2 causes of morbidity and mortality become relatively more important. The exception is sub-Saharan Africa, where Group 1 remains the dominant cause of disease burden. In the older age group, not surprisingly, noncommunicable diseases constitute the major cause of disease burden.

Injuries make an important contribution throughout, but with no very obvious linkage to overall burden or socioeconomic status.

We can use the DALY as an outcome measure for cost-effectiveness analysis. We can be particularly excited about any intervention that will save a DALY, buy a DALY, or win a DALY for less than $100 expended. That has got to be a very attractive purchase. Fortunately, we have cost-effective interventions suitable for application even in low-income countries for the top causes of burden of disease for children. For adults, the situation is not so good.

Yet—and this is key—having a highly cost-effective intervention and actually purchasing that intervention are two very different matters. For example, although cost-effective interventions are available for the top four contributors to burden among women aged 15–44 years, there is no country in which we are purchasing these interventions to the level indicated by the prevalence of the problems.

We have the interventions, they are highly cost-effective, but we do not buy them. You can visit many low-income countries and find that while STDs are not being diagnosed and not being treated, despite the highly cost-effective nature of such interventions, at the same time, public moneys are being used in the capital cities, especially in teaching hospitals, to do, for example, open-heart surgery.

We are spending public money on highly inefficient investments at the same time we are failing to purchase highly efficient investments. This is not only true of Ghana; it is also true of the United Kingdom, the United States, and all other countries. It represents a major challenge in the allocation of health resources, particularly public resources.

This type of analysis will tend to set the agenda for the future. Relative burden will change through time: HIV/AIDS will rise in women, for example, and tuberculosis will increase in both sexes. The challenge is to keep a careful eye on the numbers and to develop better—more effective and more cost-effective—weapons against these particular causes of disease burden. Having gotten them, we must then actually purchase them.

What are the challenges over the next three decades? Although we might have different lists, they will probably overlap. My list follows: First is closing the gap—the concept of equity between countries and equity within countries. In every low-income country today, there are subpopulations who enjoy a health status roughly similar to that in OECD countries. Yet other subpopulations have a health status that is orders of magnitude worse. Thus, closing gaps within countries can be a powerful motivator for politicians and communities and could also be a powerful approach for analysis. In my own country, the United Kingdom, the large gaps in health status between the folks who are disadvantaged and those who are not have at last become a matter of public debate.

Second is resisting emerging infections. HIV/AIDS is the most important and widespread, but there are others, as we are well aware. These challenges are not going to go away; HIV/AIDS remains a rampant pandemic and is now spreading rapidly through Asia with no evidence that it is going to be attenuated by weapons currently in our possession.

Third are the resurgent plagues, of which tuberculosis and malaria are the most salient. Fourth, we must confront the tobacco-related epidemic. A country such as the United States, where so much work has been done on to-bacco-related ill health and so many public and private measures have been taken to attenuate this epidemic, must bear in mind that in most of the world, tobacco abuse rates in men are rising and tobacco abuse rates in women are increasing rapidly. Most low-income countries in the world have no discernible policy toward tobacco. You can visit country after country and find no government position on tobacco, no legislation, no use of price control, no use of effective public education campaigns.

In most of the world, the epidemic of tobacco use has not peaked, and the epidemic of tobacco-related ill health will extend well into the next century.

Fifth is preparing for the new agenda in the low-income countries of the world. Cardiovascular disease, diabetes, cancer, and injuries are already major problems and are becoming increasingly prominent, primarily because of the aging of the population.

Now, let me make one or two comments about regional differences as we look toward 2020. In subdividing the world, we realize that we are not all on the same trajectory as far as health trends are concerned. It is interesting to look at countries that do much better or much worse than predicted by their income level.

A striking outlier is China, which—as we well know—is an over-achiever in health in relationship to its wealth. It is worth debating whether those achievements are robust and secure.

In the United States, we worry about having 30 million uninsured people. The Chinese have over 700 million uninsured people whose medical expenses are out-of-pocket. There are doubts about maintenance of the essential public health and preventive services that put China in its strong position. Unless the right action is taken now, the special position that China occupies may not be sustained.

Another outlier is the northern part of Central and Eastern Europe. Central and Eastern Europe in this definition includes all of the former Soviet Union and its satellites. Imagine a group of countries of which the most southerly is Hungary, the most easterly is Russia, and the most northerly are the Baltics. These countries are poor health performers in relation to their wealth.

The story there is dramatic. In the late 1950s, Czechs and Slovaks were more healthy than Austrians in terms of life expectancy. By 1990, they had become much less healthy. There, male life expectancy had declined, whereas in Austria, male life expectancy had, as with the rest of the OECD countries, improved steadily, leaving a major life expectancy gap. That life expectancy gap did not occur because of the revolution of 1989–1990; it occurred during the last four decades of the Communist regime. This pattern is repeated throughout the subregion.

Since the revolutions that occurred in the late 1980s and early 1990s, we have seen two distinct patterns. One pattern is that the gap between Eastern and Western Europe continues to grow wider because health conditions in the former Communist countries continue to decline and, in some cases, decline at an even greater rate.

Contrasting patterns are found in the more successful reforming countries of the region; here, decline or stagnation has come to an end, and life expectancy has begun to increase, although the gap has not necessarily begun to close.

What of the work of the World Bank in this changing scene? For all regions of the world, the Bank currently has a portfolio worth $8 billion, which is invested in 153 projects in 78 countries. Those 78 countries are spread across all regions of the world.

In terms of the numbers of projects, they are highest in Africa, while the largest proportion of the $8 billion is invested in Latin America and the Caribbean.

That commitment by the World Bank has grown very rapidly. Prior to about 1980, the Bank had very little activity in the health sector. The World Bank has become a major source of external finance for the health sector of low- and middle-income countries.

Investing that money wisely, getting those investments right, and designing projects in the most appropriate way are incredibly important. The Bank will have to call increasingly on the international health community and the community of experts worldwide, including the United States, for guidance on how to target those large investments so that they have the most positive effects both on health and on the efficiency and equity of health services in borrowing countries.

A dimension of this that cannot be avoided if one is speaking in this particular capital city is the IDA, the International Development Association. As some of you are aware, the loan and credit money provided by the World Bank is provided partly by IBRD, the International Bank for Reconstruction and Development, at near commercial interest rates, and partially by IDA, at zero percent interest with a 50-year repayment period—highly concessionary lending. The IDA window of the World Bank is accessible to any country that has a per capita GNP of less than about $865 per year.

Sixty percent of the $8 billion in the Bank's current health portfolio comes from IDA.

IDA, as you are aware, is threatened by current debates taking place in this city. The results of these debates will affect the pattern of World Bank support to health sectors around the world. The survival of IDA is essential for the Bank's ability to maintain a high level of assistance to the poorest countries, where of course the needs are undeniably the greatest.

In conclusion, what of science? For a given level of national wealth—$5,000, or any other example—the amount of national health associated with it is far greater today than 100 years ago. It is far greater today even than 30 years ago and will be far greater in 2020 than it is today. It would be interesting to have a 3-day meeting of the IOM devoted entirely to debating the reasons for this pattern.

The main reason, I suggest, is that, whereas 100 years ago we knew almost nothing about the etiology, prevention, treatment, and rehabilitation of disease, today we know an enormous amount. By 2020, we will know much more.

In this environment of expanding knowledge, the behavior of individuals is affected. For example, people will wash their hands in ways they would not have thought of in 1870. This knowledge also affects the behavior of households and the behavior of communities. It affects the behavior of corporate entities, and it greatly affects the behavior of states and national governments and how they choose to spend the public dollars entrusted to them.

It is this power—the knowledge environment generated by health science and by biomedical science—that has, in my view, been the main engine of health improvement in the 20th century.

My conclusions are, in light of this and of the challenges ahead, medical science matters; public health science matters; you, the teachers and researchers in the biomedical area, matter; and the Institute of Medicine matters. Long may it prosper.

DISCUSSION*

PARTICIPANT: Could you expand on your comments on environmental risks?

DR. FEACHEM: Are you referring to the Central and Eastern Europe story or speaking more broadly?

PARTICIPANT: Both.

DR. FEACHEM: Clearly, this is an ever-moving area of scientific knowledge. I think our impressions at the moment are that if we look globally, there is one subset of the environment that matters hugely and all other factors matter relatively little. The factors that are globally important are the domestic environment, water that is clean and plentiful near the home, removal of human waste from the home, indoor air pollution, and drainage. These things have a huge effect on global human health.

Environmentalists who comment on Central and Eastern Europe like to believe that the environment, particularly air pollution, is a major cause of some of the negative health trends that I discussed, but the evidence is to the contrary. The evidence is that environmental factors are a minor cause of those health trends. That is an unpopular view in Central and Eastern Europe, because it is much more comfortable to believe that somebody else's problem, which causes the smoky chimneys that pollute the air I am forced to breathe, is responsible, rather than my body mass index of 35, my cigarette smoking, and my poor driving.

PARTICIPANT: Can the World Bank do more to raise the profile of diseases such as diabetes in developing countries?

DR. FEACHEM: Indeed, there is a debate going on at the World Bank, and between the Bank and its clients, about the health agenda over the next decade. With countries such as India the topic of noncommunicable diseases is clearly on the table. Among other client governments there is a variable degree of awareness of—and interest in—the noncommunicable diseases at this time. One of the roles the Bank can play, aided and abetted by expert

---

*Because of technical difficulties, the participants' questions had to be reconstructed; therefore, they are not verbatim.

opinion such as that from the IOM, is to put these topics firmly on the table and include them in the debate so that some actions may follow.

PARTICIPANT: Does the World Bank listen to the opinions of professionals and people at large, or only to the voice of governments?

DR. FEACHEM: As you are aware, the Bank has a new, dynamic, and outward-looking president in James D. Wolfensohn. This is high on his agenda—to improve both the actuality and the perception of the Bank's interaction with all stakeholders in the health sector and other sectors in countries in which we are lenders. We have to deal formally with governments, but we are aware of and interested in broadening the debate to include all those able to contribute to our common goal of improved health and more equitable and efficient health services.

Part II:
Implications for Health

Introduction to
Afternoon Session

John M. Eisenberg, M.D., M.B.A.
Chairman of the Department of Medicine, Physician-in-Chief,
and Anton and Margaret Fuisz Professor of Medicine,
Georgetown University Medical Center

T he year 2020 will bring with it changes in the health care system and the way in which we care for our patients that will be greatly influenced by other events in society and other changes that surround us.

As I thought about this, I was reminded of Oliver Wendell Holmes when he said, "The truth is that medicine is as sensitive to outside influences—political, religious, philosophical and even imaginative—as is the barometer to changes in the atmospheric density."

This session is concerned not only with how the barometer of medicine is likely to respond to changes in the atmospheric density of the health care system, but also with the kinds of changes that will occur within the health care system itself.

One of the major changes, of course, is the movement toward a more market-driven health economy. Our image of Sir William Osler as the hero of American medicine slowly fades into the portrait of Adam Smith. I must confess, as one who spent some time at the Wharton School, that I have worshipped at the altar of the free market and I continue to believe that it is remarkably powerful. The market brings us certain strengths that help us to respond to the public's choices and desires in ways that are remarkably quick, breathtaking, and sometimes disruptive. It has caused us to question, especially during the past three years, what the proper role of government is in a market-driven health economy.

As we look to the year 2020, we must ask continuously about the proper role of government in such a market-driven economy. Will the market be governed by government? Will government try to make the market work as well as it possibly can, to make up for information deficiencies and to help consumers to choose more wisely? Will it ensure that there is entry into and exit from the market that allows the kinds of choices to be made among options that otherwise might not be available if entry were limited? If the government's role is to help the market work well, what will happen when the market cannot work—when access is limited and certain people cannot get in? What will happen to education and to research—areas in which the market generally fails us?

Third, if government is to help ensure that the failures of the market are compensated either by making the market work as well as possible or by substituting for the market, there will also be a role in which government is the prudent purchaser from the market, whether we are talking about Medicare, Medicaid, the Department of Veterans Affairs, or the Defense Department. The role of government in the future will be an important one, even with a market-driven economy.

Much of the discussion in this session traces the boundaries between a market-driven economy in health care and the proper role of society as it is expressed through government. I anticipate learning about how that might help our institutions evolve. What might our institutions become in the future, as horizontal and vertical integration proceeds, as networks continue to be developed, as risk sharing leads to the need for information sharing, and as our relationships change? The latter include relationships between clinicians and their patients, relationships among different kinds of clinicians, and importantly, relationships among patients, clinicians, and their organizations.

In closing, let me offer an observation that I think is very helpful for those of us who are so much of this market and of health care that sometimes we cannot understand what is happening. Walter Lippman once said, "You cannot see the play and be in it." In this session, we hear from some people who are in the play, but who seem to be able to see it as well.

Cardinal John Newman once said that he did not have much use for all of this futurism: "I do not ask to see the distant future. One step is enough for me." However, in the current health care environment, one step might not be enough.

Risk and Responsibility: The Evolution of Health Care Payment

Jeff Goldsmith, Ph.D.
President, Health Futures, Inc.

As you might surmise from the title "Risk and Responsibility," this discussion reflects a deep personal interest in the issue of risk and how we as individuals in society manage it. That interest is more than an intellectual interest; it is a visceral one. In my professional work, I think about how changes in science and technology and in the organization and financing of care are going to change this trillion dollar health care activity that we are all a part of.

In my personal life, I train in Tae Kwon Do and ski off cliffs and out of helicopters. I believe that my professional activities are more dangerous than my recreational activities. This topic, the issue of how health care financing is going to change, is fraught with particular danger because so much of the beta, the variability in our financing system, is accounted for by political and normative forces that are notoriously difficult to predict.

As a person who has been doing this for a while, I have discovered that there is a great deal of difference between forecasting 25 years ahead and skiing off a cliff. In skiing, the more air you put under your skis, the more danger there is. In forecasting, the more air you put under your forecast, the less danger.

I discovered this about 10 years ago when I was asked by *Hospitals* magazine to write an article on the U.S. health care system in the year 2036. My first reaction to that request was, I suspect, not all that different from the

---

This talk is based on material that originally appeared in an article coauthored with M. Goran and J. Nackel, "Managed Care Comes of Age" in *Healthcare Forum Journal* 38(5):14–24, 1995.

way anyone would react, "Who knows?" Then, I realized that not only would I be dead in the year 2036, but so would every single person who had read the article. It was a liberating moment, because I realized that I could say whatever I wanted and no one would know how wrong I was.

Now, with the year 2020, I think we are in a slightly more dangerous situation, because there is the prospect that a few of us will be alive in the year 2020. Christine Cassel, a gerontologist and a friend from the University of Chicago (who is now at Mount Sinai in New York), gave me these prevalence forecasts for Alzheimer's disease suggesting that the survivors among us probably will not remember what was said. So it seems worthwhile taking some predictive risks.

It seems to me that the central direction of change in our financing system in health care in the last 15 years, which I see persisting for the next generation, is the shifting of economic risk and responsibility for health care costs from the government and employers to health plans and the health care provider community, and subsequently to individuals. I see this trend accelerating in the next 15 years as high-risk populations enter managed care pools. Of course, the growing number of uninsured people in our country fits this trend because what we are seeing there is a shift in economic risk from the government and employers, who are trimming their responsibility for certain segments of the employed population, to individuals and to a lesser extent, to the health care institutions that have responsibility for them.

The idea of shifting economic risk is the central theme of this discussion—how we manage that shifting of risk and how an important societal institution, namely managed care, is going to change as it assumes a steadily increasing amount of risk—the risk involved in an aging population.

Even though managed care has been with us as an institution for the better part of 60 years, it is my contention that it is still in its adolescence as a societal form. Three powerful forces are operating now in the health care financing landscape that are going fundamentally to transform that business as it moves forward into the first part of the next century.

Fifteen years ago there were only about 10 million people enrolled in HMOs (health maintenance organizations) in the United States, and that population probably represented the healthiest 10 million people in the country. The typical HMO subscriber was probably white, a blue- or white-collar worker with a family comprised of people who did not get sick very often. The largest costs in a typical health plan were obstetrics and pediatrics.

Clearly, a major change that is taking place in the composition of the managed care population at risk is the flood of high-risk individuals entering that pool. These individuals come from the two highest-risk subgroups in our population: the multifunctionally impaired, chronically ill elderly population, for whom Medicare is currently responsible, and the poorest of the poor—people enrolled in Medicaid programs—who bring all of the societally driven risks associated with their socioeconomic status into the risk pool.

So if managed care 10 years ago was about managing the care of people who did not get sick very often, managed care in 10 or 15 years is going to be about trying to manage the risk of the sickest people in our society.

The second major change is the change in the competitive structure of health insurance itself. Ten or fifteen years ago, managed care plans had the luxury of pricing their product under the spacious cost umbrella provided by completely unmanaged indemnity and Blue Cross plans that basically wrote checks after the fact for decisions made by professionals and institutions.

These forms of health care payment were primarily giant sluiceways for other people's money. If the Blue Cross plan in your area or Metropolitan Life raised its rates by 20 percent a year, your managed care plan could raise them by 15 percent and look like a hero. That has also changed dramatically. Health insurance premiums, at least in the private sector in the United States, are falling, and falling most rapidly in those communities with the highest managed care penetration. How long this continues is anyone's guess, but disinflation in the private health financing system is a very new and welcome trend.

So we have gone from a situation in which managed care was able to acquire a lot of additional revenue while caring for relatively healthy people, to an environment in which the per capita amount of dollars going to the plans is shrinking in real terms.

The fact that indemnity health insurance based on after-the-fact payment is disappearing creates a tremendous challenge for managed care plans to invent a new rationale for their existence besides "we are simply cheaper than conventional health insurance."

The third change in health care financing, which I believe is the most significant, is that we are moving from an event-driven to a risk-driven health care payment system. One could make the argument that until recently, managed care was not really about managing care at all. It was what I call event-driven cost avoidance (see Figure 1), which means trying to minimize the cost of caring for someone after the person has become sick by

EVOLUTION OF MANAGED CARE

Stage I—"Event-Driven Cost Avoidance"

Principal Objective: Reduce hospitalization after illness commences

Subsidiary Objective: Slow specialty physician use

Mechanisms: Queuing
Preadmission review
Concurrent review
Outpatient/inpatient substitution
Provider discounts

Figure 1.

reducing the amount of hospital care used and interposing bureaucratic barriers to access to practitioners. In this model, the health care system is arrayed at the bottom of the cliff waiting for people to arrive, while the managed care plan tries to reduce the cost of cleaning up the mess after the patient has crashed onto the rocks.

This model is rapidly failing as a viable definition of the managed care business, particularly in the many communities in our country where managed care has been established for a long time, because the first major layer of savings that managed care tries to achieve—reducing per capita hospital use—is nearing bottom in a lot of places. This is certainly not the case in the East, where a variety of regulatory forces, along with resistance from the medical community and from the labor unions, have retarded its growth.

In my home town of Portland, Oregon, inpatient hospital use for the entire community, including the elderly, is about 370 days per thousand people and falling. We used to argue in health planning about whether four hospital beds per thousand was the right standard. Managed care plans in my home town are running at 0.8 bed per thousand, including the elderly, and falling, 70 percent of the Medicare population have enrolled in managed care plans voluntarily.

The other thing that is happening to managed care in communities where it has been long established, which should not surprise anyone who thinks about the nature of price competition, is that price is becoming a less and less useful guide to selecting a health plan. In addition to premiums falling, what is happening in most of these communities is that the difference in price between health plans is narrowing to the point of perhaps a 5–7 percent difference between the least expensive and the most expensive plans.

As if that were not enough, if price does not serve to differentiate health plans from one another in these highly competitive markets, the networks that provide services under different health plans are becoming increasingly identical. So on two important points, price and access are far less useful than formerly for consumers and employers to decide which plan they want to choose. Thus, in addition to exhausting the easy savings from reducing hospitalization, managed care plans face an additional challenge—how to differentiate themselves from one another.

What you heard from the Jackson Hole Group this summer is that managed care is in the process of inventing a new rationale for its existence. That rationale has been called value improvement (see Figure 2). I believe that over the next 7–10 years, the activities that we have listed here are going to be the predominant preoccupations not only of managed care executives, but also of their medical directorates and the physicians who participate in them.

There are really two challenges in value improvement. Challenge number one is to try to control the resource intensity of the use of clinical services across an episode of illness, while also trying to improve consumer family satisfaction with the clinical experience. As has already been pointed out, this issue of improving the value of services is a data- and information-intensive enterprise. Trying actually to understand how costs are gener-

EVOLUTION OF MANAGED CARE

Stage II—"Value Improvement"

Principal Objective: Control resource intensity

Subsidiary Objective: Improved consumer satisfaction

Mechanisms: Capitation of specialists
TQM/CQI
"Right-sizing"
Clinical pathways
"Patient focus"
Outcome monitoring
Controlling units of service:
   —Drugs
   —ICU
   —Lab tests
   —Diagnostic tests
   —Surgical procedures

Figure 2.

ated—how they are built up at the bedside, in the clinic, and in the physician's office—requires orders-of-magnitude better information about what clinicians actually do in treating patients, as well as systems that enable you to examine the variation in patterns of care across a particular type of clinical problem.

The most important issue of all, the issue that Dr. Jack Wennberg at Dartmouth and Dr. Robert Brooks of the RAND Corporation have been working on, is how to reduce the huge variation in resource consumption by practitioners for treating the same kinds of problems. There is obviously a tremendous amount of gold to be mined here as managed care plans seek, with the assistance of professional societies and their colleagues in academic medicine, to rationalize clinical decision-making—to create what has been called, in Don Detmer's presentation, an evidence-based framework that defines what constitutes best clinical practice for a patient presenting with a particular constellation of health risks.

This is going to be a very difficult challenge, whose goal is to be able to say to the consumer or to the employer—with data to support it—that our plan is doing a better job of utilizing your clinical dollars than the other health plans available to you.

Now, it is worth asking ourselves what happens to this business when you have removed the excess hospital utilization, as well as a lot of the variation in clinical practice that does not necessarily add value in terms of patient outcomes. I have a somewhat unconventional answer to that question—an answer driven by the logic of managing risk itself. Managed care plans are going to discover that they are really in the public health business. So, to respond effectively to pressure from the large numbers of high-risk

EVOLUTION OF MANAGED CARE

Stage III—"Health Improvement"

Principal Objective: Population-based health status improvement

Subsidiary Objective: Reduced health services cost

Mechanisms: Pooled risk capitation
               "Prediction and management" model
               Risk appraisal
               Targeted intervention

                  —At-risk individuals/families
                  —Community/environmental factors

                "Case management"
                Multidisciplinary teams

Figure 3.

people who will be migrating into managed care plans, improving the health status of defined populations ultimately is going to be what differentiates health plans from one another. These differences will provide essentially a noneconomic rationale for deciding which health plans to use.

The idea that market forces may be pushing private health plans in the direction of a population-based approach to health improvement is counter-intuitive. Familiar tools of public health, such as epidemiological modeling and forecasting, as well as the somewhat more invasive relationship between public health systems and the people they serve, will provide a guide for private health plans and participating physicians as they move into the next century.

What this represents is movement from an event-driven to a risk-driven framework for health care payment. Instead of diagnosis and treatment as its principal business, our health care system will have to predict health risk and try to manage that risk before it flowers into illness and cost. We will get a lot of assistance from the extraordinary developments taking place in both genetics and immunology. These advances will enable us to identify much more precisely the embedded genetic risks in populations and in individuals, while blood tests will identify the degree to which those risks have begun to flower into illness (see Figure 3).

I think the technological capacity in the year 2020 will be sufficiently powerful to enable us to frame the risks that an individual runs at any point in life—at almost any point beyond conception—and to present a challenge to that individual and that family, as well as society, to create the best outcome framed against that risk. What the health system will try to do is provide information about health risks to individuals and create a user-friendly framework that enables individuals to collaborate with the health system to manage risk before it flowers into illness.

Many questions are raised by this progression—a price-driven to a value-driven to a health-status-driven health care payment system. You can see (Figure 4) that the focus of the health care system shifts from the hospital to the physician network to the risks embedded in the community itself.

The most important shift, however, is on the third line—the locus of control. Many cost management activities taking place in our health care system today are superimposed upon the doctor/patient relationship. Physicians find it necessary to have phone conversations with a far-distant nurse about what they can or cannot do to patients they have known all of their lives for any clinical decision that involves more than a few hundred dollars.

As health plans shift risk to physician organizations, responsibility for effective clinical decision making is going to devolve from a bank of nurses talking over 800 numbers to the communities of physicians who are at-risk. Much of the superstructure that has been built up to prevent physicians from abusing the fee-for-service system will become unnecessary. The real challenge will be for physicians in organizational frameworks that range broadly from independent practice associations to group practices. As risk devolves onto these physician enterprises, so will responsibility for setting clinical standards and policing them. Thus, we see the responsibility for intelligent, thoughtful, conservative clinical decision making devolving back onto the physician communities that are at-risk.

As we move into Stage III, we see a further devolution of responsibility onto individuals, as well as a tremendous societal debate over the appropriate division of responsibility between families, society at large, and the health system for the task of health improvement. The fact that 40 million people in the United States have no health insurance means that those 40 million people are 100 percent at-risk for the costs and consequences of their medical problems. This is obviously inappropriate at many different levels. Yet the idea that we also have individuals—in many cases, wealthy individuals—with no economic risk or responsibility for the cost of care is, in a way, equally intolerable to our society.

I think that many issues are raised by how we arrange things in our society, but that as citizens, a balance is needed between our right to health care and our responsibility to use the information available about our health risks to minimize them through our behavior.

There is no question that an increasing amount of our health is going to be mediated by the decisions we make as individuals, as well as by decisions made by society about how to organize our communities. There is something fundamentally unbalanced about the idea that we have a right to health care without a reasonably rigorously defined set of obligations to live in a healthy way. It seems to me that as we move into this third stage, we are going to have a dialogue about what we as citizens should be doing to manage our own health risk and the appropriate division of labor between the health system and individuals or families in managing that risk.

EVOLUTION OF MANAGED CARE

	Stage I	Stage II	Stage III
Objective Function	Price	Value/consumer satisfaction	Health status improvement
Cost Targets	Inpatient days	Resource intensity	Health risks
Locus of Control	External	Peer driven	"Contract" with family
Focal Point	Inpatient hospital	Physician network	Home/neighborhood

Figure 4.

If there was a "hole" in the Jackson Hole paradigm, it was that the sole role of the consumer in their model was to select the less expensive health plan. That role does not seem a sufficiently active one to enable us, particularly during an epidemic of chronic disease, to use effectively the societal tools—the science and technology—that will be available to help us manage the risk. So when I refer to a contract with the family, I am not sure where that contract will evolve or be developed. It may well be that the appropriate locus for dialogue about the balance of risk and responsibility between society and the individual is first of all, the health plan.

What is really going on in Congress right now, is that the federal government and state governments are getting out of the business of paying hospitals and doctors and into the business of delegating risk and responsibility for future costs to managed care plans. They are also delegating the very messy tasks associated with the transition in our health care system that will occur as managed care increases in influence; the reduction in the disparity of income between primary care physicians and specialists, and liquidation of the huge surplus of bed capacity and technology that we have in our health care system. These issues are also being delegated to health plans and, unfortunately, so is the issue of defining the appropriate relationship between the individual and the state, and between the individual and the community.

One could perhaps argue that there is a fourth stage that does not appear in Figure 4. Stage IV is probably what has gone on in Oregon. There are very real moral hazard-related limits on how much we can expect health plans to do in the way of limiting access to clinical services, particularly for the terminally ill and the handicapped. There is a very real limit to how much people will be willing to tolerate managed care plan involvement in the hospice movement or in the right to die a dignified death. If these issues are resolved by the narrow economic interests of the health plan, there may be tremendous tension between individual wishes and the economic interests of the plan.

The issue of an appropriate allocation of health dollars across a community and a region is going to be begged by the growth in managed care and will eventually result in a community dialogue that, with hope, will address them in a systematic and broad-based way as was the case in Oregon. Are we getting the maximum value for our societal investment in health care services?

The purpose of this discussion is not to minimize the costs or the difficulty of this transition, but to point out that there are opportunities involved in the idea of managing health risk as a central activity in our health care system. It is an integrative idea and something that competing institutions will not be able to do by themselves. I think we are going to see collaboration among competing health plans, and between private health plans and the public health sector. I think we are going to see bridges built between the health establishment and other institutions in our communities that are implicated in the economic risk associated with illness. These include our educational system, our criminal justice system, and our family and social services networks. To accomplish this transition in an intelligent and thoughtful way is going to require building such bridges. It is going to require creating a virtual community to support our country's health.

Risk and Responsibility: The Evolution of Health Care Payment— Response

John M. Eisenberg, M.D., M.B.A.
Chairman, Department of Medicine, Physician-in-Chief, and Anton and Margaret Fuisz Professor of Medicine, Georgetown University Medical Center

I feel as though Dr. Goldsmith has brought us to a Tom Cruise film festival. As I think about his comments about the role of medicine in modern society, *Top Gun* comes to mind—high technology at work. We are doing the best that we possibly can, and we are delivering. Yet he also seems to describe a world of health care that is like *The Firm*: that is, we are working for an organization that is incredibly unscrupulous but we have not figured it out yet. The future of the health care enterprise seems to be so uncertain, according to Dr. Goldsmith, that it sounds like another entry in the film festival—*Risky Business*.

As I think about the future of health care, this idea of entering a world of risky business is a frightening one. I remember the adage that 10 percent of the people incur 70 percent of medical care costs. If that is true and if we the providers and the patients are going to be responsible for decreasing 70 percent of the costs that are incurred by 10 percent of the people, the question is, "What do we know about where those costs are coming from? Can we discern the risks so we can alter them?"

I came into medicine at a time when the most popular books about health care were written by Ivan Illich, Thomas McKuen, and Rick Carlson, who said that health care providers did not make any difference. We might as well close up shop because we were having no impact on health. It is very reassuring to note that pundits now believe that we make such a tremendous difference in the outcomes of care for patients, and the costs, that we have so much control, that we should assume all of the risk.

Yet perhaps that is not what Dr. Goldsmith really means. Perhaps he is saying that there is much variation in what we do and that, since we aren't sure which parts of what we do make any difference, society is going to give the risk back to us, to deal with in any way we want.

No, I do not think that is what he really means either. Yet, it is of concern at a time when we do not have good measures of risk stratification, when we know that measures of severity of disease have lagged behind what the market demands, that we are being told—those of us in academic medicine, the Agency for Health Care Policy and Research, NIH, and the entire health enterprise—that we have really failed to provide the public with the information it needs to decide what kind of health care it wants.

This gets back, of course, to the question that has been a recurring theme throughout our discussion: Where is the information that we can apply in accepting this responsibility for the risk and variation in health care costs and outcomes?

There is a wonderful line from T. S. Eliot in which he writes, "Where is the wisdom that we lost in knowledge, and where is the knowledge that we lost in information?" He wrote that early this century, before there was any idea of a silicon chip. The meaning was that information alone will not do it—that we must take the information in all of these databases, create knowledge from it, and from that create wisdom. I hope we can serve the public as well as possible in that regard.

Let me turn to a second theme in response to Dr. Goldsmith's comments. I am fascinated by the idea that risk would be shifted to providers and to the public. If that is the case, the public will probably ask, who are these third-party payers and what have they done for me lately? Why am I paying overhead to these people who then pass my money on to doctors and hospitals and ask them to take the risk? In some ways, we are led to the same conclusion as one major entrepreneur in health care, who said recently about his hospital chain, "We will be the K-Mart of the health care system." In essence, we will be the wholesalers. We do not need these retailers. We do not need these intermediaries.

As I think about Borders moving in and shoving out the little mom-and-pop bookstores, and about Giant and Safeway moving in and displacing small grocery stores, it is not just a matter of accepting risk but a matter of accepting size, integration, and networking and what they mean to the health care system. I think, in general, that the results will be salutary. Yet, as a physician who relishes the intimacy of the primary care relationship, I wonder what that size and the assumption of risk will mean to the doctor–patient relationship. Is there a way that we can accept that risk and yet retain the intimacy of our relationship with our patients?

I think about calling up my travel agent for an airplane ticket. I do not relish the idea of my travel agent working for an airline any more than I like the idea that my doctor has large incentives to do either more or less than is appropriate for me. I would like for us to develop ways in which we could assume risk, adopt integration and networking to take advantage of size, and yet retain that intimacy and trust that is so important in health care.

Let me finally ask a question about the assumption of risk, whether we as a society—not just one health care system or another, or doctors, or small groups of physicians—will assume risk? The more we learn about genetic predispositions and genetic markers for disease, the more we learn that risk is not merely whether or not you drink and whether or not you smoke. It is not just whether or not your physician has advised you about healthy habits. It may be uncontrollable, essentially the hand we were dealt.

If we reject Thomas McKuen and Ivan Illich and assume that we can make a difference, we still are left with the fact that the people who come to us are in very disparate risk categories. There is the jeopardy that we will reject as our patients people who are BRCA positive or people who have positive family histories of disease without even knowing what genes are involved, figuring that some gene is lurking that increases the risk of our patients' incurring costs. This is going to test our community and our societal values; it will also test the ability of members of the health care system to share responsibility for one another.

DISCUSSION

PARTICIPANT: We had, I would say, a rather optimistic picture of the future, at least from our Dr. Goldsmith. I wonder if there is not more to the dark side when I think about the fact that the major for-profit concerns have their stock market value determined by the ratio of their promotion and administration expenses to their expenses on actual health care—the higher the latter, the lower the former. Are we headed in quite the opposite direction, where we will not get to Stage III or Stage IV but will end up in a swamp unless we make it impossible by law, I suppose, for nonphysicians to practice medicine? That is a very eschatological and unhappy view, but it seems to be a view that we perhaps ought to hear a little more about.

DR. GOLDSMITH: Well, to be rigorous about that, their stock price rises and falls in accord with per-share earnings momentum, not the loss ratio. Let's be clear about what the market is valuing.

PARTICIPANT: No, but I understand that market analysts use that figure as a shortcut.

DR. GOLDSMITH: Well, these folks are my colleagues. A lot of them are poised right now over the sell button on their computers because they already know what many of the people who are worried about for-profit medicine dominating health care fail to remember, which is that health insurance is a brutally cyclical business. On the downside of that cycle, you do not want your capital tied up in an enterprise that is losing money on underwriting medical risk.

I do not know how many of you saw the market response to U.S. Healthcare's missing its earning by a penny or whatever it was. The company lost 40 percent of its market value. A lot of the money in for-profit stocks, which I follow closely, is "hot money." As Mexico and Brazil have taught us, you cannot really build a solid foundation on hot money. When the hot money leaves, the executives of those firms are going to have to ask themselves if they can create a noneconomic rationale for their subscribers and for the physicians that work with them to continue participating in their system. So perhaps a bloodless University of Chicago person sees slightly more market discipline than someone who is looking narrowly at this allegedly inexorable trend of the gobbling up of not-for-profit medicine.

DR. HELMS: Bob Helms with AEI. I just wanted to get you to talk a little more about how this market will evolve over time. I do not disagree with your view that managed care may play itself out, but I wanted to ask what the market response will be to traditional fee-for-service medicine—if it exists at all now.

Dr. Eisenberg has talked about the concept of physicians becoming agents. Is it likely that the market will demand more of that as people want more information? It seems that in a certain sense, physicians have the knowledge base to become those agents. I wonder if there is going to be a market reaction to managed care? Is the market waiting to take advantage of it?

DR. GOLDSMITH: That is an interesting question, because I see two distinct responses to the shifting of risk on the part of physicians. Response number one—from perhaps 80 percent of the medical communities in most places—is a flight from risk into salaried employment by hospitals and health plans. Physicians are essentially saying that the business risks associated with their continuing to be individual entrepreneurs are unmanageable economically or psychologically, and they want shelter in a large enterprise.

What I have been saying to my health system CEO colleagues is that the risk is shifting to them. I know enough about large health care organizations to know what happens when you get up to $1 billion or $2 billion. The person who is really at risk is the CEO, and that individual cannot push the risk down far enough into the organization to get the kind of behavior change on the part of professionals on the frontlines, who are actually getting blood on their shoes, to really affect the cost or value of services.

The remaining 20 percent, which is just now beginning to be visible, is the flight toward risk on the part of physicians who are organizing a variety of risk-bearing, risk-seeking enterprises. They are saying, "Give us a percentage of the premium, give us the responsibility, and we will figure out how to organize medical care for the population that we represent." This is the most exciting thing happening in medicine right now—even more exciting than a lot of the outcomes research and evidence-based medicine—that is the organizational response to the shifting of risk.

I really believe that physicians have the power to manage this transition. If they throw the power away—if they give it to Columbia HCA or the

Massachusetts General Hospital—obviously these organizations will take it. We are seeing in California and elsewhere that physicians are able to organize risk-bearing enterprises successfully and to accept the risk and responsibility, not only for the economics, but for the health outcomes as well.

DR. HELMS: Are these responses in conflict—the idea that a small group practice of three or four people does not want to accept risk, but that the same people are willing to accept risk if they are part of a larger corporation?

DR. GOLDSMITH: Well, it is not necessarily a corporation. Some people have given the independent practice association (IPA) up for lost; that is not an accurate read. Many private practice physicians have found what might be termed "virtual relationships" to one another that enable them to bear and manage risk without becoming part of a large medical bureaucracy. A lot of people looking at the medical care marketplace right now are saying that the Mayo Clinics, the large regional group practices, are going to dominate. I think they are wrong. There are diseconomies of scale and coordination in medical practice that are going to severely hamper the 800-person or 1,200-person physician organization because it just cannot get the discipline or achieve the degree of change in values and behavior necessary to manage risk in an effective way.

DR. SHINE: Where does the capital come from?

DR. GOLDSMITH: There is no shortage of capital. The Mayo Clinic is sitting on, in round numbers, $2 billion. So is Kaiser. There is money lying around all over the place. I do not see capital as the constraining element here. This is not a capital-intensive business; it is a knowledge business.

In terms of the issues that Dr. Detmer has raised—of how much money it takes to create an information system in which people can talk to one another and access databases—vendors will tell you that hundreds of millions of dollars are needed. If we pay close attention to what Dr. Detmer has said, it may not be capital that constrains us as much as the ability to create the connections and to achieve what is a more difficult challenge—professional consensus about appropriate standards of treatment. That is where the real gap is. I do not think capital constrains physicians to the degree that people marketing capital or systems would have them believe.

PARTICIPANT: One of the key words in Figure 4 is "contract." What kind of innovative work is going on in the area of contracts and contract development as you move toward Stage III? It is not just the contract between the purchaser and the managed care organization, it also involves the managed care organization with the provider and possibly with the patient. As I look at the kind of services we are going to need, I am worried about that contract link with the managed care organization, which in immature markets may be selling lives every three to four years as they move in and out.

DR. GOLDSMITH: Well, this is an area that I am raising as a policy issue now but do not see people formally addressing yet. Political scientists talk about a social contract—that is really the kind of contract that I mean.

A University of Chicago person would think about the insurance contract as a mix of incentives, a set of signals that society is sending to people about what it wants them to do. It seems to me that we are going to begin redesigning that contract to reward individuals who make health-conserving decisions and to place some economic risks in front of individuals who have a measure of control over the amount of health risk they are creating for themselves.

Dr. Eisenberg is correct. As we get more and more information about genetic risk, more and more questions are going to be begged by the role of the individual agency in promoting that risk and flowing into illness. The insurance contract by itself is a blunt instrument in moving that risk, but we at least ought to be trying to move the margin. We should be telling people through the insurance contract that we want them to make conservative, thoughtful decisions in managing their own health risks. I am not sure how long it will take before people think about this. The $50 deductible or $5 copay for prescriptions is not what I am talking about. We are going to have to decide what behavior we want to invest with some degree of economic risk. We must figure out how to protect individuals with limited resources from having too much risk placed on them. We have to begin thinking about what we want people to do to help improve their own health.

PARTICIPANT: You say very little about the role of government in this evolution, other than that it is shedding risk and will continue to do so. Surely there has to be a major role for government to ensure that the market will work as you envision it. Could you speak to that?

DR. GOLDSMITH: There is a huge role. I do not know how many of you saw Matthew Miller's piece in the December 11, 1995, issue of *New Republic* on the Medicare program. It is an absolute bafflement to me why we should use cost-based payment to pay HMOs to take care of Medicare recipients, which is essentially the framework we use now. We have built up this huge body of cost that varies threefold from community to community, and we now give health plans a license to "mine" that variation instead of, as Miller suggested, putting the contracts out to bid and letting health plans competitively bid for and accept the risk for the Medicare population.

On the issue of the loss ratio, one of the reasons HMOs are really nervous right now is because that loss ratio is an enormously tempting regulatory target. I could see a slightly different administration, a somewhat less risk-averse, more courageous one than the one we have in office now, saying that it is not going to contract with a health plan that spends less than 90 percent of the premium dollar on health services. I think that government has an enormous role to play here; I am just daunted by how difficult it is for our federal government to make rational decisions in health financing.

DR. EISENBERG: Let me add something to that. The idea of contracting for managed care organizations and having bids was described by Miller in that *New Republic* article as having come out of the Physician Payment Review Commission (PPRC). So I must state that it was developed beautifully by the PPRC staff and is described in last year's annual report. We will see if anyone decides to take it up. It is a risky but interesting idea.

DR. JOHNS: I am Mike Johns from Johns Hopkins. I am interested in understanding how the marketplace is going to accommodate the ever-expanding numbers of uninsured. Although there seems to be a sense that shifting government health dollars to the states will allow them to spread those dollars further, I expect that more people will drop off the roles and there will be an increasing number of uninsured. Somebody ultimately bears the cost of that, generally the middle class in some way.

How will this system deal with that? Will there be any sense of accountability? For example, these physician groups that you talk about—the new entrepreneurial physicians that we see coming together—will they be willing to take on the responsibility for some of these populations? Will anybody be able to afford to take on those responsibilities? Who is going to step up to the plate?

DR. GOLDSMITH: The system that we have now gives academic health centers and urban public hospitals the quasi-governmental responsibility of taxing the rest of the health care system to pay for services provided to the uninsured. I agree that there is nothing in the current round of "reforms" that is going to do anything other than increase the number of people who are not covered. I think this is fundamentally irresponsible, flawed social policy.

You could argue from a strategic point of view that the federal government is now the principal driver in health cost inflation in the United States. President Clinton's spirited defense of a double-digit rate of increase in public spending for two public health financing programs is not adding anything to the debate over how to get more affordable care for the population that is uninsured. I do not see the private sector leaping forward to take responsibility for these folks, and if public costs continue rising at the present double-digit rate, we won't get an affordable federal or state response to the problem of the uninsured.

Yet I do think there are opportunities to pool the purchasing power of individual and small group health insurers and give them the same kind of per capita cost advantage that large employers enjoy. We are not going to get very far in solving this problem without revisiting the health policy debate. With hope, we will not waste as much time as we have in the last two years.

Institutions and Health

Lawrence S. Lewin, M.B.A.
Chairman and CEO, Lewin-VHI Health Group

My topic this afternoon is institutions; however, it is important for me to say that I have a few qualifications. I am going to focus rather narrowly on institutions. In doing so, I am mindful that many other dynamics are at work in the health care field that I will not say a great deal about.

Like most revolutions, managed care is a powerful movement. Its ultimate consequences are unpredictable, which makes predicting the future difficult. In doing so I am reminded of a history professor I had in college, who, when asked if he would predict whether democracy would come to Eastern Europe said, "You want me to predict the future? Are you kidding? I have enough trouble predicting the past."

I cannot even try to predict where we will be in 25 years, in large measure because where we will be in 25 years depends very much on the responses of institutions in the nearer future. So my focus is more on what I think will be happening in the next 5–10 years; then perhaps we can speculate a little on where that takes us in the year 2020.

Let me begin with a brief summary of the major new forces that will affect the evolution and survival of health care institutions. The first factor is the ascendancy of the purchaser. In a recent conference sponsored by the Robert Wood Johnson Foundation, Lynn Etheridge and others talked about how we are headed for a period of competition that is largely a matter of purchasers' getting control from insurers. This theme comes up throughout my remarks.

The second factor is the shift in emphasis via capitation from providing medical services to sustaining health—from managing care to managing health. I want to make a number of comments about this. It has been said, and I agree, that we are not lacking the knowledge of how to do this. We do in fact have most of the technology and information that would enable us to help people keep and sustain their health. The problem is that the current system is poorly organized to fulfill this goal. Our institutions are neither motivated nor capable of doing this effectively.

The third factor, which is a bit of a sleeper and may put me at odds a little with Dr. Goldsmith, is the importance of capital in building competitive systems. I believe capital has an important role, and the availability and distribution of capital will have an impact on which institutions are able to survive.

These factors will provide a tremendous challenge to many institutions but none as much as academic medical centers. These venerable institutions will be particularly challenged by almost a Malthusian kind of process and will need to reinvent, reengineer, and "right-size" (or downsize) themselves in major ways.

We are in a very dramatic period. Things are changing multidimensionally every day. Part of the challenge that faces many institutions is a schizophrenic one of dealing with a world we know is moving in a different direction while we still have to survive in the current world.

Let me now talk about some major changes affecting institutions. I want you to be aware not just of the nature and direction of these changes that are occurring, but of the schizophrenic situation in which many providers or institutions find themselves in dealing with these shifts.

The first change is with respect to payment. We have moved from a fee-for-service environment to one influenced by DRGs, and now to one markedly increasing capitation. Because capitation represents a 180-degree change from fee for service, many institutions are wondering what to do in the process.

The second is the kind of outcomes that are desired. In the past we have expected people to provide treatment for sickness. Now we also are concerned about outcomes, which is to say the outcomes of care applied to a particular event or episode of illness. We are moving, as a result of capitation, more and more in the direction of the sustenance or management of health.

The way providers have measured desired outcomes in the past is to simply count the services delivered and be paid for the number of services provided. The new system is dominated by price—who can offer a benefit package at the lowest price?

In the future we are headed toward a system in which buyers will be seeking value. Value is quality divided by cost. The problem with value is that most people assert that they provide it but offer very little proof. As a consequence, when you say to most purchasers of care that you provide high quality or high value, they do not pay much attention because they feel that providers themselves do not know either relatively or absolutely how much value they are providing.

Another major change is the growing influence of purchasers at the expense of providers. Insurers, however, will maintain some leverage as well. Here I want to take issue with one point of Dr. Goldsmith's—the trend of risks flowing downstream from insurers to providers. If you look at what is happening in California, the insurance companies who now have a lot of market power are downloading risk, but very selectively. They are doing so because they understand clearly that if you can manage against risk, you can produce enormous savings, and they do not want to give up those savings.

What we have told our clients is that there are primarily three things they should think about. (1) The bad news is that risk is coming to providers. (2) The worse news is that you may not be able to get risk when it does come because insurers will retain it. (3) The worst news of all is that you may get risk but not be able to manage it effectively enough to avoid losing your shirt in the process.

Another major change relates to mission-driven institutions, that is, those that provide services for which they get reimbursed but that also have other interests such as charity care or education. These institutions are today in a major crisis. Government is no longer willing to pay for their missions, and the market certainly is showing a decreased willingness to bear these cost subsidization burdens. Therefore, there is a serious question as to where the money for mission financing in the form of disproportionate share payments (e.g, indirect medical education payments) is going to come from in the future.

The final change is with respect to investment capital. In the past, institutions have been able to raise investment capital because they were using it to build facilities; these are easily collateralized and so institutions were able to get debt financing. Today, the things institutions need money for are less easily collateralized and require equity financing. Entities that do not have access to equity markets or do not have large endowments will be at risk.

I propose that provider institutions are going to have to take on this era of the purchaser in four ways: The first is that they are going to have to gain market power. They must do this to become indispensable to their markets.

The second is that they are going to have to develop more effective means of integrating hospitals and physicians and related physician networks.

The third is the need to internalize managed care in order to sustain health. Much of what has come to be thought of as managed care is an imposition of very blunt instruments external to provider institutions. I believe that in the future, successful institutions will have to internalize these processes if they are going to be able to document outcomes and manage health.

Finally, there will be a need to develop new sources of investment capital. Let me provide some details on these four points.

Gaining market power. If you remember nothing else, keep this important point and the one that immediately follows it in mind. Institutions that do not gain market power effectively will not be forces 25 years from now; they will be historical relics. Gaining market power is a positioning strategy designed to increase or regain influence with purchasers who will be the major buyers.

Institutions will need to gain control in several respects. The first is volume of services—access to covered lives; this simply means having the market share to produce the income necessary to support the institution, particularly when the teaching mission is involved.

The second is referral patterns. Being part of a system that has a lot of covered lives is not worth much if you do not see the patients needed to educate your students or to generate enough income to sustain the institution. In bargaining with third parties, whether they are purchasers or insurers, the more bargaining power you have, the more you can lock up referral patterns. Part of the bargaining process is establishing price—prices that are high enough to support margin.

How does one establish this kind of bargaining power? Well, a word that Bob Derzon and I thought up a few years ago was "indispensability." Making yourself indispensable in your community means that third parties will have to deal with you in an effective way.

Indispensability is relative to the market and local conditions. There is no one formula. A couple of factors are worth noting, however. The first is having sufficient size to be taken seriously and sufficient product differentiation for people to come to you rather than to others. The latter means that others will not be able easily to substitute their services for yours.

The second is competitive pricing along with cost control. Obviously, all other things being equal, and, in many cases, even if they are not equal, those who get to play will be those who can offer a competitive price. However, a competitive price does not mean much if you are not able to control your costs within that price. So the ability to control costs is absolutely essential.

The last factor is access to and control over primary care managers. I use the term primary care managers, rather than primary care physicians, advisedly. I do not believe that primary care managers need necessarily be primary care physicians. Many others could be trained to play that role, and they need not be physicians. Particularly as we move more and more toward managing health rather than just managing care, nonphysicians will play an increasingly important role.

Now, there are a lot of theories about the form that institutional structures might take. The most popular one, of course, is horizontal or vertical integration. In my view, neither horizontal nor vertical integration, aside from the need to be a certain size, is essential to achieving increased market power or indispensability. At times, in fact, these can be counterproductive.

The acid test in any merger or consideration of a partnership should be whether it will make you more indispensable according to the criteria I have just described. In that regard, while much is made of economies of scale, if you are going to make an effort to integrate either horizontally or vertically, it is important to ensure that real cost reductions and downsizing are achievable and to agree on them in advance. Some enterprising researcher at the Federal Trade Commission went back and looked at various institutions several years after mergers had been approved that promised great savings, and found that virtually none of the savings promised in order to avoid antitrust violations had materialized.

What our firm has been recommending to our clients is that they build such agreements into the letter of understanding or the final document. Although that makes the merger process more complicated, to achieve real savings there must be some agreement up front.

In a recent paper, Dr. Goldsmith pointed out some reasons it is so difficult for hospitals to downsize. Even when mergers take place, they do not downsize very frequently—often because of internal constituencies. The future is going to demand significant downsizing. Studies we have done project reductions of 30–50 percent in inpatient care. Institutions that do not effectively downsize are going to be doomed in part because of the cost burden, but also in part because they are at such a disadvantage in trying to negotiate with insurance companies.

It is quite possible that ownership, which is viewed as an increased form of control over cost and over quality, could actually be counterproductive for purposes of downsizing and achieving innovation.

When our firm takes a preliminary look at a proposed merger of two hospitals because of the need to downsize, I sometimes think about the late Arthur Okun's quip about the two hikers confronted by a grizzly bear. Says one hiker to the other, "I don't have to outrun the grizzly bear; I only have to outrun you." Instead of trying to prop up this other institution by merger, why not simply let it fail so that you can have its patients? I think this is the way the market dynamic is beginning to work, although there is much resistance to it. The point is that if you are going to merge to improve your position in the market, you must have a clear agreement that downsizing will occur. Increasingly, we are finding merger candidates recognizing that they have to do just that.

Now, another strategy for institutions is to integrate upstream, or to integrate backward, and offer insurance products. Indeed, Medicare legislation proposed in both the House and the Senate would confer some advantages on provider organizations in actually becoming health plans and offering their services. These provisions are potentially very attractive for moving in this direction.

Intermountain Health Care (IHC), on whose board I serve, is one the first systems to do exactly that. I have been able to see the tremendous benefits that have accrued. Interestingly, the benefits are not only economic. IHC gained a strong market position. It is making money on its health plan and is protecting its market. The need to manage care effectively to compete in a very competitive market has forced IHC to bring physicians to the table and find ways to manage care more effectively. This requires managerial skills and financial reserves that most providers do not have. Indeed, the financial requirements to sustain risk and offer an insurance product can be a severe drain on capital that perhaps could be used for better purposes.

Another issue that tends to come up in these discussions is the issue pertaining to broad geographic coverage. Some of our clients in New York, for example, are trying to build broad networks, acquiring as many hospitals and physician groups as they can, under what I believe is the benighted assumption that this will give them a stronger position in the market.

There is, however, another point of view here that needs to be understood if institutions are going to respond productively rather than just follow the latest fad. That is, insurers, HMOs, and other bulk purchasers are not necessarily drawn to institutions that have broad geographic coverage because they can often "cut a better deal" on their own. They would view an all-or-nothing package as a disadvantage. If that is the case, why go to the effort to build a broad package in which everyone has to be included?

Obviously, strength in the market is an issue. If you are going to integrate backward and become an insurance plan, it does make a difference. If the rest of the country begins to move in the direction of the Twin Cities, with direct contracting by purchasers with providers, that kind of a system might make sense. However, in my view, too much is being made of broad geographic coverage. Such a strategy, if not well thought out, can become a tremendous consumer of management energy and capital.

The next point is the importance of integrating caregivers and facilities effectively. Purchasers are going to favor organizations that can integrate financial accountability and clinical autonomy. This is a critical point; it is the essence of Dr. Goldsmith's earlier comments with respect to taking risks. The point is that up to now, most financial accountability has been imposed externally, and has been a very sore point with physicians in particular. If physicians are going to regain their clinical autonomy, autonomy that they now find threatened by outside institutions, they will have to embrace financial accountability; otherwise, others will assume that role.

In the future, physicians will be less autonomous as individuals. Indeed, the solo practice or small group practice of single-specialty physicians is probably a thing of the past. As that happens, physicians who are bright and entrepreneurial will form new organizational structures to seek economic and managerial power. Along the way, many of them will find themselves responding to the offers of for-profit, investor-owned companies. I believe that in the process of doing so, they will get a fair amount of money and be able immediately to monetize their income stream. However, as these systems form they are going increasingly to become counterforces to effective integration of services. Over time, therefore, these arrangements probably will not remain robust and will begin to disintegrate. Unless these organizations form other strategic alliances that I believe are necessary—they will not by themselves become a major force for going forward. As an aside, hospitals, and even academic medical centers, will also be responding to some of these for-profits. I discuss the capital part of that shortly.

Hospitals will have to redefine their notions of governance and economics if they are to really create a common destiny with physician groups. This is more than just a virtual versus a formal system. It has to do with allowing physicians to play a role in governance, as well as a share in the gains of the ongoing enterprise.

The next point has to do with internalizing managed care or achieving managed health. Here I think we get back to the point raised earlier about risk sharing and information sharing. Capitation basically rewards systems of care that can sustain health. Why hasn't that happened right now? Well, first of all, we do not have a lot of real capitation. Secondly, the issues of risk selection and turnover have removed the rewards for institutions to attract and manage people with the chronic diseases that are most susceptible to health maintenance or health sustenance. As we solve these problems of risk selection and turnover, I believe that capitation will become a powerful engine in moving us toward sustaining health.

As I have said, we are very poorly organized to do that. Providers are not rewarded for keeping people healthy. The payment system does not support that absent capitation. Few HMOs have therefore seriously pursued it, making their money instead on provider discounts, control of volume, and risk selection.

Few physician groups have the personnel, the information systems, or the financial incentives to do it, but there are exceptions. Friendly Hills is one example of a 150-person physician group that has developed many internal protocols designed to sustain health. Interestingly, it arrived at those protocols largely by looking at what high-cost items were beginning to eat into its profits.

The priority of sustaining health and health maintenance is probably going to await a time when purchasers really demand outcomes data. At that time we will be faced with a scarce supply of primary care managers—people who are trained in helping people change their habits and in caring for people with chronic conditions that need more careful monitoring.

The real leverage here will come from the wider use of telecommunications technology, telemetrics, and telemedicine, which I have for some years believed to be the most underrated technology in health care. This technology will become absolutely critical if we change the paradigm to one of managing health—enabling us to reach the population in ways that support both educational monitoring and compliance with treatment regimens.

New sources of investment capital are important not so much because there is not enough capital in the system, but because the capital is distributed in biased ways. The way capital is distributed in the future will have a major impact on which institutions will survive.

Meeting the demands of purchasers and responding to all of the things I have mentioned relating to leverage in the market, indicate that for the time being, insurers may have an advantage over providers, given their access to capital and the systems and capabilities they have developed over time. This may change in the future. New capital requirements cannot rely on traditional debt instruments. Some new instruments are needed. We are probably going to have to turn to the investment banking community to come up with some alternatives, particularly for not-for-profit groups to finance reserve and other system requirements for provider service organizations, as well as for information needs.

Not all of these need be dramatic, "big-buck" kinds of strategies. They may be efforts that some associations can mount without the help of investment bankers. Whatever they are, if we are to move beyond debt financing secured by facilities, the investment community will demand greater focus and strategic preparedness on the part of benefit institutions, in a word, a lot more discipline.

Finally, we probably will be seeing a whole range of interfaces and new partnerships that will merge investor-owned and not-for-profit institutions. I think that this will be a surprise to many of us.

Now, some specific comments regarding academic medical centers. The challenges facing medical centers are obviously major. Academic medical centers will need to redefine their mission and structure in order to coordinate clinical and academic enterprises. Within many academic institutions there is tremendous tension between the clinical and the academic worlds. Many chairs are still academically oriented, interested principally in research and teaching, although they preside over departments that are required to meet the needs of the market and are not particularly oriented toward doing so.

There is a need to reorient both the clinical and the academic enterprises to achieve indispensability. This will require changes particularly within the academic community to support these clinical imperatives. It will also require some rethinking of the directions and shape of both teaching and research programs to see whether they should be coordinated with the new imperatives necessary to achieve indispensability in the market. This will require a very close look at the centers' teaching programs to see whether they are focused and to determine whether their research portfolios are consistent with the strategic orientation of the whole academic medical center.

Obviously, moving away from tertiary and quaternary care to promoting wellness, and the financial incentives this requires, will be enormous challenges. They probably will not be accomplished within the academic medical center alone, but will require partnerships with other groups.

Information systems must not only manage health but also have the ability to demonstrate value. You must be able to have systems that support the notion that you are doing a better job than others.

Downsizing capacity is going to be critical. To some extent it will be necessary on the clinical side, but it will be especially true in teaching and research.

There will have to be a reincarnation of academic departments, which I believe generally are poorly organized for the purposes that the future will demand. First, academic departments must be reincarnated as parts of group practices, rather than the silo system that currently exists. Then, the programs offered to the public must be market responsive, which means that they probably will have to be interdisciplinary—a real challenge for traditional academic medicine. This will be a major change, and some schools are already taking it on.

Finally, there are the historical, often pernicious town–gown conflicts that many academic enterprises will have to address as they seek to become community providers. The arguments on both sides are solid and have enough merit not to be ignored, but they must be overcome.

To meet all of these challenges, academic medical centers are going to have to develop structures that are strategic, decisive, yet constant to their important mission and core values.

The traditional academic structure has been characterized by one of my colleagues as based on the thousand points of veto. That is quite appropriate in many academic settings, but it is not very helpful when you are trying to respond to the market.

In some cases, these new realities will mean separating clinical enterprises from the university and from their state ownership. In many cases we may see a trend toward separation of the clinical parts of the academic medical centers from the medical schools, which will be less compromised in their mission. In any event, there is going to be a lot of change.

Academic medical centers are going to have to deal realistically with the leaner diet that major public and private purchasers have ordered for them. Purchasers are increasingly unwilling to pay the premiums they are asked to pay to support medical education. In that regard, it is important for academic institutions to recognize that Americans and their governments have other pressing priorities to support; therefore, their dependence on government for support is going to be limited.

Two more points in summary. To succeed, academic institutions are going to have to abandon the culture of paternalism and arrogance that makes it so difficult for them to form partnerships with lesser parties. I say that not so much to be critical but as one who has great affection for, and many clients among, academic medical centers. Nevertheless, academic medical centers must face up to this statement because it is true.

Finally, I am not prepared to condemn academic medical centers as dinosaurs. I believe they have tremendous advantages that can put them in strong positions. They perhaps have the greatest ability to create the technologies that will make a real difference in the nation's ongoing endeavors to combat serious illnesses. Close relationships already exist between physicians and hospitals. Academic medical centers command high regard and loyalty in their communities as flagship institutions. They have strongly dedicated to values. Many of them even have large endowments. By facing the challenges of trying to increase their market power, academic medical centers can go a long way toward surviving in the new environment.

Institutions and Health: Response

Robert M. Carey, M.D.
Dean, James Carroll Flippin Professor of Medical Science, and
Professor of Medicine, University of Virginia School of Medicine

My charge is to present a personal vision of health-related institutions 25 years from now and to project the resultant changes in health. I shall confine my remarks to the institutions I know best: academic health centers.

As Mr. Lewin suggests, academic health centers are especially vulnerable to external forces in a rapidly evolving health care system. The system is struggling to balance the demands of employers and insurers, the needs of the public, institutional traditions, professional values, available resources, market pressures, and consumer choices.

In many ways, academic health centers are anomalies within market-driven managed care systems. Managed care anticipates that the majority of health services will be devoted to routine patient care in a relatively healthy population. Managed care systems benefit directly from medical education, cost-effective technologies, and innovations in patient care. Under the current structures, however, they do not pay for them, nor do they pass on the costs to their customers in the form of higher premiums. In this respect, managed care systems at the moment are "free riders." These managed care systems and other market forces are making us look anew at some of the essential components of the health care system and who will fund them. Managed care begs the question, "Who will bear the costs of education, the development of new knowledge, and the care of the poor?" Certainly some of these costs were assigned inappropriately to revenues from clinical care in the past. How will academic health centers respond to these challenges, and how will they deliver health care in the year 2020?

INTEGRATION OF HEALTH CARE SERVICES

Academic health centers will be organized and operated so as to provide integration of health care services. At present, most academic health centers deliver health care in a fragmented manner. Medical schools and teaching hospitals have separate organizational structures that foster duplication and asymmetry of work processes, lack uniform service quality, have uneven and often competing incentives, and entail little or no alignment in a shared vision. Inpatient services usually are supported by the hospital; outpatient services, by the medical school clinical departments. Physician care is provided by these departments along traditional medical disciplinary lines. The clinical departments function as independent financial centers, usually with high priority on the departmental bottom line. The organization as a whole receives less attention. There is insufficient resource sharing for the benefit of the entire institution, and decisions entail a multiplicity of steps and opportunities for veto.

In order to address market demands for cost-effective coordinated health care services, academic health centers will have to revise their approach. *They must place the patient at the center of attention.* Institutions will have to ask themselves, "What are the patient's perceptions each step of the way?" Academic centers will have to develop integration of the various components of health care centered on the patient. Physicians who are members of the medical school faculty will have to work collaboratively with their nonphysician colleagues, especially in nursing and administration. For example, program-oriented interdisciplinary service centers (e.g., cancer, cardiovascular disease, neurosciences), involving physician and nonphysician staff, could be created. Service centers would be oriented to the best way to deliver care to patients. In this model, the physician (medical school) and nonphysician (hospital) leaders could share responsibility (and possibly budgets) for contracting all in- and outpatient care within the service center. Physician and nonphysician teams could be created to ensure collaboration throughout the organization.

The service center would be responsible for standards of care, collaboration in care, resource allocation, program themes, and physician manpower. Under this model, academic departments would continue to exist, but health services would be organized around interdisciplinary service centers supported by the departments. In some instances, there may be complete congruence between department and service center; in others, multiple departments will be involved in a service center. Details of organization and governance will differ among academic health centers. I predict that most of the centers that survive will have created an integrated clinical enterprise with smooth interface among the academic departments, their clinical practice plans, and the teaching hospitals. Academic health centers will function more as unified entities, some with full asset merger, in the care of patients.

COMMUNITY OUTREACH

Academic health centers will extend themselves outward to patients and populations. Many academic health centers have insulated themselves as "ivory towers" or have become "courts of last resort" for complicated or severely ill patients. The patient was placed in the position of having to visit many different subspecialists, each focusing on different aspects of the same problem. Under managed care, subspecialists will be encouraged to *move toward the patient.* They will become involved early in the care provided by the generalist. Generalists and subspecialists will work as teams. The team will try to prevent a subspecialist office visit or to recommend either the most appropriate, cost-effective applications of specialized technology or hospitalization. In the ambulatory setting, doctors will see patients in one location whenever possible. This will lead to maximally efficient, effective, and timely care.

Successful academic health centers will work with other network providers to respond to a diverse set of health needs that move beyond the walls of the traditional academic health center. The framework will have to be expanded to contain two legitimate and equal foci of attention—the patient for whom we care and the population for whom we are responsible. The needs will include wellness, nutrition, hospice care, home visits, home monitoring, home therapy, and rehabilitation. Advances in technology will create new and cost-effective ways of shifting from inpatient to ambulatory to home-based care. Home chemotherapy and the teaching hospice will be legitimate activities of academic health centers. Academic health centers will be concerned with the health of the entire population they serve. Together with their networks, academic health centers will concentrate on prevention. Centers also will contribute to the assessment, mitigation, and prevention of the impact of environmental hazards on health.

Academic health centers will extend themselves to patients and populations by telemedicine, health informatics, and other efficient methods of communication. These methods will be used for consultation, patient information transfer, patient self-learning, home-based care, continuing medical education, clinical outcomes research, and analysis of the cost and quality of care. In these practice areas, academic health centers will make major contributions to all health networks, especially in medically underserved inner city or rural areas.

NEW HEALTH CENTER PARTNERSHIPS

Academic health centers will engage in new partnerships in health care. Successful academic health centers of the future will develop partnerships: (1) Partnerships with generalist physicians, because they can help preserve patient flow; and (2) partnerships also with diverse elements such as nursing homes, pharmacies, health departments, free clinics, pharmaceutical companies, and information technology businesses. Academic health centers will be viewed in a dispersed model. They will be seen as dynamic partnerships with other elements of the health care system. These new linkages will better serve the public need. They will open the door for new opportunities, especially in health services research oriented to population health. The academic health center will change from a relatively self-centered, independent academic model to a "cooperative culture." It will recognize that interdependence is a higher order of principle than independence. Centers will become full partners on behalf of the health needs of society.

SOME CONCLUSIONS

Academic health centers, as well as the health system at large, are struggling with the medical, economic, social, political, ethical, and legal issues surrounding the current revolution in health care finance and delivery. Observers perceive the changes and their future implications in disparate ways. At one end of the spectrum, those longing for the "good old days" in medicine envision the health care system of tomorrow as sterile. They foresee large integrated networks seamlessly moving patients through "vertical and horizontal layers of care" in a cost-effective manner. They predict strict regulation of the use of pharmaceuticals, diagnostic tests, clinical services, practice patterns, and benefit packages. One imagines a kind of Orwellian production line of health care. Under this model, practitioner and patient widgets represent industry outputs, with competing private insurers fueling the enterprise. The emphasis would be on the system of care at the expense of the individual.

At the other end of the spectrum, free-market economists observe the current maelstrom in health care as a market self-correction. Competitive forces will be wringing out waste and demand from a supply-heavy commodity. They envision a transformed health care system with balance in physician supply and levels of services regulated by individual consumer demand. The power of the individual purchasers—the patients, or their employers, or corporate proxies—would ensure quality and cost-effectiveness. Unfortunately, under this scenario, many of the value-added participants in health care (e.g., medical researchers or educators) would decline as the invisible hand of competition squeezes the profits and subsidies that sustained them. Without these investments, the entire enterprise would suffer or perhaps fail.

The health care of the next century will probably fall somewhere in the middle of these extremes. It will be neither totally automated nor market driven. The current cross-subsidies and patchwork of services to assist those outside the mainstream of insurance, I predict, will eventually drive us to national health insurance.

Will we be a healthier society? There is every indication that we will be healthier as we understand the genetic and epidemiologic basis of disease and incorporate preventive strategies for the public.

Will we have someone to whom we can turn for compassionate care? The "soul" of health care is the physician–patient relationship. This must be preserved as we experiment with organizational structures and practice patterns. Ethical conflicts, moral tensions, and financial concerns arising from conflicts of interest should be between physicians and administrators, *never* between physicians and patients. To be sure, physicians must meet the needs of society to reduce and stabilize health spending, but our highest calling is to respond to the needs of our patients and, indeed, of mankind.

DISCUSSION

DR. SMITH: Dave Smith, from Texas, where I'm commissioner of Health. I may have missed a question earlier—about the model of community-oriented primary care, which we have been tinkering with for some time at Parkland in Texas. Do you see a potential for resurgence and some interest in this model that describes a lot of what we have talked about here today?

MR. LEWIN: Yes. Community-oriented health care has been an important part of all of this. I think we are seeing a great acceleration of interest in it as Medicaid moves more and more toward managed care. Although the absence of an appropriate financing system has been an obstacle, that by itself is not going to solve the problem. We need technology. We need people trained in that.

The only caution I would offer is that sustaining health in those populations probably will be the greatest challenge of all. The failure of those organizations to be effective in managing health should not be taken as a reason not to extend community-oriented health care elsewhere because the poor and the elderly represent the greatest challenge. That is certainly a major issue.

DR. BLUMBERG: Baruch Blumberg from Philadelphia. From what we have heard from Congressman Porter, the discussion of the future of the academic health centers, and the total market orientation that is proposed, there seems little future for scientific medical research and, in effect, a declaration of resigning from that post. That may decrease scientific arrogance in this country, but it would also take us from the position of leadership that we have enjoyed for the last several decades. With the disappearance or decrease in academic scientific leadership, technology—which so much is dependent on—will be late in coming.

DR. CAREY: There is no question that the academic missions of academic health centers are under threat, both the teaching and the research missions. However, there is some hope. First, in an integrated model academic health centers can pool all of the resources available and make them available for research and teaching. Second, creativity needs to be exercised in decisions regarding the flow of funds from the clinical toward the academic enterprise—funds not only from revenues, but also from increased efficiency in patient care can come into the equation. Under these circumstances, significant financial resources can be plowed back into the academic mission.

In addition, I cannot overemphasize the importance of educating the public on the value of biomedical research so that a substantial constituency will be behind the NIH and other research funding sources.

MR. LEWIN: I do not think for a moment that this country has lost interest in scientific inquiry. What we are hearing is that the level of resources available requires a very careful analysis of whether the resources being consumed are in fact productive. The suspicion is that there is room for more efficient use of those resources. I know this is always a problem in dealing with science. I think that there is a great deal of unsponsored research that probably is justified more by the tenure of the faculty rather than the value of the research. At least, that is the perception.

DR. FEACHEM: I was very struck that Mr. Lewin's and Dr. Carey's presentations did not contain a single reference to any other country, which I thought was quite notable. In terms of some of the recurrent themes—purchasers, contracting, capitation, and academic medical centers—it did strike me that a country I know well has quite advanced purchasing authorities in both the private and the public sectors. They are funded through weighted capitation, with which there is much experience. The contracting functions are now fairly well advanced and elaborated.

In terms of academic medical centers, having run one for a number of years, I can attest that rampant democracy does not reign and that there are not a thousand points of veto in the way the United Kingdom runs its academic medical centers. In making these contrasts, I do not want to particularly draw attention to the United Kingdom. I want to make a more general point about international comparisons. Lest it be thought that the United States is perhaps insufficiently comparable to some other OECD countries, it is good to bear in mind two factors: First, most of the rest of the OECD buys better health status for its population at about 8 percent of gross domestic product (GDP) than the United States does at 15 percent of GDP.

Second, compared specifically with the United Kingdom, the amount of public funds that the United States invests in health care, both in dollars per capita and as a percentage of GDP is larger than the amount of public funds that the U.K. invests in health care. So your publicly funded system is larger than the U.K.'s. I think international comparisons can very much fertilize these debates.

DR. CAREY: I agree.

DR. LEWIN: We have much to learn. I think much of what other countries are learning from the United States they learn by observing some of our failures. Although some of the techniques that we are now beginning to move toward may be instructive and helpful to some other countries that are moving away from purely publicly funded systems.

New Knowledge for Health

Baruch S. Blumberg, M.D., Ph.D.
Fox Chase Distinguished Scientist, Fox Chase Cancer Center

The assignment given to me is to discuss new knowledge related to health in the 21st century. This is a very broad topic and I have decided to consider one aspect of it, genetic susceptibility to disease and the maintenance of good health.

The genome project is on schedule and it is likely that all, or nearly all, of the human genome will have been identified and sequenced early in the next millennium. Francis C. Collins, director of the National Center for Human Genome Research, has estimated its completion in 2002 or 2003. Many genes that "cause" inherited diseases have been identified. In addition, a large number of *susceptibility* genes are already known and many more will be identified in the future. Susceptibility genes affect the probability of incurring an illness and are often associated with exposure to an agent in the environment. For example, susceptibility genes for several cancers have been identified; BRCA I and BRCA 2 for cancer of the breast, AT for cancer of the breast and other cancers, p53 for a large variety of cancers, and many others.

There are also genetic effects on exogenous and environmental agents that affect the host and are potentially pathogenic. For example, genes have been identified that control the production of enzymes that detoxify carcinogenic or other disease-causing agents. These enzyme systems are often polymorphic, that is, they are controlled by a series of alleles segregating at the same locus, which can impart different levels of activity. Some of the alleles may produce levels of enzyme that detoxify the exogenous agents, while other alleles will control enzymes that are null or are deficient in their ability to detoxify. Hence different individuals will respond very differently to the same level of exposure. Those with increased inherited susceptibility, that is, who have inherited the defective detoxifying enzyme, are at greater risk of becoming ill after exposure to the exogenous agent than those who have inherited an alternate allele with adequate levels of enzyme. This information can be used to discover exogenous agents that increase risk for a particular disease. The allele that determines the lesser-detoxifying enzyme would occur in higher frequency in those diseases for which the exogenous agent in question increases risk. This technique could focus attention on the elimination of these dangerous elements from the environment. I will provide an example of this phenomenon later in this chapter.

The rapid progress of the Human Genome Project opens the possibility of the identification of potential disease in clinically normal individuals, and its prevention before it becomes symptomatic or life threatening. This raises the prospect of living a disease-free life, until the inevitable demands of the aging process lead to death at an advanced age. Yet this happy goal is achievable only if the pace and sophistication of research are increased.

The notion of preventive medicine at a personal level is in some respects antithetical to the traditional social contract between a physician trained in the Western tradition and his or her patient. The contract reads, in essence: "If you become sick, then I am available for help. If you are not ill, that is someone else's responsibility." It will be interesting to see how this doctor and "nonpatient" relation develops in the coming decades.

I would like to digress a moment from the discussion of prevention and talk briefly about therapeutic medicine. The science and art of medicine are dependent on accurate diagnosis. The physician attempts to sort patients into diagnostic categories or "bins." Once classified, the patient is assigned a prescribed pattern of therapy. Diagnosis will become more specific as the genetic and other etiologies of more and more symptom complexes become known. Instead of treating a patient as a member of a *category,* each patient will have his or her *unique* diagnosis and treatment. Oddly, this pattern is similar to the style of some non-Western indigenous medical systems in which individualized treatment is prescribed for each patient.

Currently, there are person-to-person differences in response to treatment, such as age and gender, that are known prior to the initiation of treatment and that influence treatment. With increased knowledge of the genome, more genes that influence response to drugs will be discovered and understood. The science of "pharmacogenetics" will make it possible to design therapy based on prior knowledge of the response expected.

Application of this genetic knowledge has and will raise many ethical questions, which can only be alluded to in this presentation. The organizers of the Genome Project have, wisely, included a study of actual and potential ethical problems as an integral part of the research project. This may be the first organized effort incorporated in a major research program to identify potential ethical issues before, or at the same time, that they arise. It can be said to be an anti-Frankenstein's monster program; a disaster-prevention strategy.

A goal of a program of identifying susceptibility genes would be to advise people at risk of the hazards, in order to mitigate them. The geneticist would also want to advise people who do not have the susceptibility genes that they are at a lower risk. However, this raises problems. To identify the presence of susceptibility requires the development of appropriate genetic probes that identify the mutations. The genomes are often large, there may be many mutations, some of them rare, and the distribution of the mutations may vary greatly from population to population. Therefore, if one or a small number of genetic tests are used and found to be negative, the individual may still be at risk. Assuring the individual to the contrary could raise medical, public health, legal and ethical problems.

Another ethical problem arises when susceptibility is detected, but there is no available intervention. Does a person want to know of an unhappy fate if prevention or cure is not possible? There are potential questions related to health and life insurance. Insurance companies may want to know who is at genetic risk in order to either exclude them from coverage and shift their care to other organizations, or to charge a higher premium. However, many of the traits we are discussing are genetic polymorphisms, which implies that there are both positive and negative values. Whereas a gene may increase risk for some diseases, it could concur advantage in other circumstances. Polymorphisms are common and any individual may carry a complex combination of advantageous and disadvantageous genes. If he or she is to be penalized for one trait, then fairness requires a search for other traits that may compensate for the disadvantage. If there is a polymorphic locus segregating in a community, then some individuals would be at greater than average risk, and others at less than average risk. Would they be offered a lower premium?

What I believe will happen is that the principle of mutuality, on which insurance premiums are often based, will prevail. A uniform fee will be charged on the understanding that the risk in a large population will even out from person to person and some intermediate risk estimate will lead to an overall economy.

Single-gene therapy has, so far, been limited in its results. This, in large part, is a consequence of technical problems, that is, the delivery of the genes into the host, duration of effect of the introduced gene, and so on. Yet, when these difficulties are surmounted others are likely to remain. Only a relatively small number of diseases are a consequence of single-gene effects, and even in these the repair or replacement of the affected gene may not cure. It is important to keep in mind that genetic control of characteristics and diseases, that is, the phenotype, is complex. Diseases and characteristic are often *polygenic,* meaning that several genes act on the phenotype. Genes are often *pleomorphic,* a single gene may have multiple effects. As we have already discussed, many genes are *polymorphic*, that is, there may be minor sequence differences at the same locus which have profound differences on the resultant protein and on pathology and physiology. Further, posttranslational events can affect function, tropism, and the quality of the protein produced by the gene. In most cases disease is a consequence of complex interactions of many genes and exogenous agents in the environment. Often it will be more expedient to intervene to change the exposure to an environmental agent than to change the genes. I will illustrate this point later.

We are frequently reminded that we live in an information age, and that the storing, distribution, and management of data are as important as its collection. This will be particularly true as knowledge of the complexity of the genome and its protein products increases. Continuing development of software to manage these data will be required before it can be fully utilized. Progress in medical research and applications will be dependent on the establishment of intellectually independent groups operating within a medical or research environment, so that people who are expert in the design and use of the data systems and also thoroughly conversant with biology will be available. As more and more scientists grow up with computers, it will be possible to find these talents within a single person. (Film lovers with an eclectic taste may recall that Buckaroo Banzai was such an individual.) More likely it will require close cooperation between scientists skilled in one or the other of the disciplines.

There will also be an increased need to construct mathematical models of biological systems. As more and more complex systems are considered, multiple genes, multiple products, multiple external agents all interacting will be chaotic if not organized into an understandable model which can then be subjected to experimental and observational testing. The current attempts to understand "complexity" in a wide variety of disciplines is a wholesome move in this direction.

It will be possible to "mine" data bases. This metaphor carries many implications; for example, the probing of data bases to obtain information that one, a *priori,* did not know was needed. An additional application is likely to come in storing data on individual human genetic "fingerprints" that carries with it ethical and legal problems. It is likely that prediction on the basis of an individual's genome will be far more complex than at appears at present when the identification of all of the genes is incomplete. Yet, there will be a large amount of predictability incorporated in an individual's genome. How will individuals react to knowing their unique fate?

The past few decades have made it apparent that we can expect new plagues and the return of old ones. When cholera first emerged from Asia into Western Europe, it traveled at the speed of a man on a horse or in a sailing ship. Infectious agents, using the vector of jet aircraft, travel at nearly the speed of sound. As the world becomes more integrated, economically, politically, and culturally, we become more exposed to infectious agents from foreign parts. Strains of bacteria, viruses, and other pathogens which, in the past had remained within geographic bounds, now travel as widely and as rapidly as their hosts. That includes pathogens that have developed resistance to antibiotics and vaccines.

In order to effect some measure of protective control over these agents, it will become necessary to establish surveillance stations, often in hitherto remote areas, to determine when new strains develop. Ideally, it would be better to know about the emergence of new pathogenic strains *before* they cause an epidemic. New strains can be identified by the determination of their gene sequences. Very detailed epidemiology and transmission histories will be

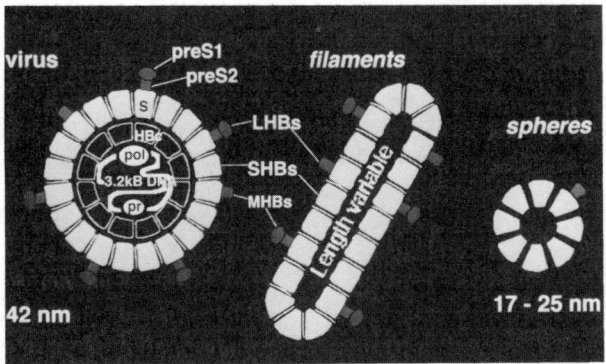

Figure 1. The three particles associated with hepatitis B virus. The large particle contains DNA and polymerase, is infectious, and can replicate. The smaller particles contain only the surface antigen. The locations of preSI and preS2, Large, Medium, and Small surface antigen proteins (HBs) are indicated.

feasible, since a unique strain from the individual who was first infected could be traced through contacts. It may be possible to determine exactly who transmitted an infection to another.

The management of this vast amount of data and the comparison of sequences to establish homologies and differences, polymorphisms, mutations, interacting genes, and other complex phenomena, will require very sophisticated data base management. New technical applications will be needed to facilitate the transmission of this information between the scene of the mutation to the scientific centers where these data can be collated and decisions on interventions made.

I have raised these issues in a general manner, but I would now like to provide concrete examples from the research with the hepatitis B virus (HBV).

The human hepatitis viruses, of which at least seven (designated HAV to HGV) are now known, are among the most important pathogens that invade humans. Each of them is a different virus, and many come from completely different families and genera, but the clinical outcomes and means of transmission are also similar for many of them. My illustration will apply to HBV, but the same principles may apply to other hepatitis viruses and other diseases.

The three particles of HBV are shown in Figure 1. The large particle is infectious and replicates. The small sphere and the rod do not have DNA and contain only surface antigen. There are about 300 million carriers of HBV in the world. Many of these people carry infectious and replicating virus, and a portion of them are at high risk of developing life-shortening chronic liver disease and hepatocellular cancer (HCC). HCC is one of the most common cancers in the world, particularly in Asia and in sub-Saharan Africa, where there are a very large number of HBV carriers. Fortunately, an effective vaccine, invented in our laboratory in 1969, has been available for more than a decade, and it has been used very effectively in national universal childhood vaccination programs. This has markedly decreased the number of carriers

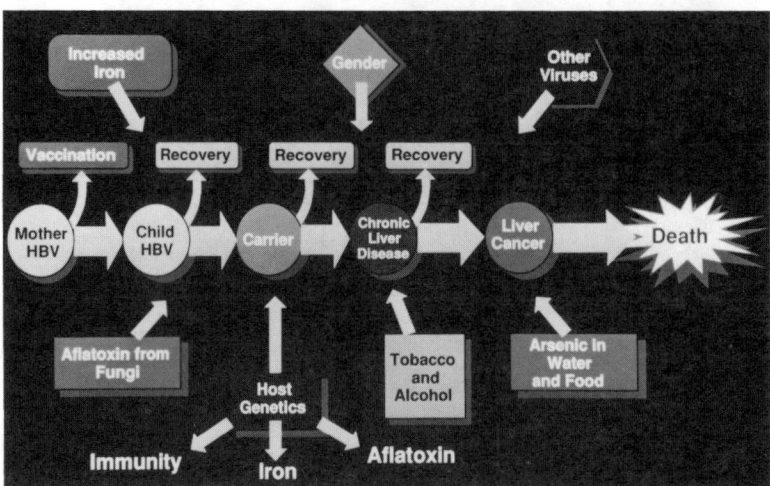

Figure 2. Diagram illustrating the possible courses of chronic HBV infection and factors that influence progression. See the text for further explanation.

among the younger age groups (i.e., up to about 10 years of age) in several places, including Taiwan, Japan, and Korea, but the number of adult carriers who were unprotected in their youth are a major medical burden and will remain so for many decades. Research on treatments of carriers of the virus and those afflicted with the disease is very encouraging, but it is important to increase our understanding of the complexity of infection and pathogenesis.

Figure 2 demonstrates the complexity of the factors that bear on infection and disease. It illustrates the course of chronic hepatitis B infection from the time of transmission in early life until the development of chronic liver disease and hepatocellular carcinoma. In Asia, in particular, children are often infected from their mothers, and this increases the probability of becoming a chronic carrier. Only a portion of those chronically infected proceed to disease and a shortened life span. What are the factors that affect the speed and direction that the infection will take? The diagram illustrates some of these, arranged in no particular order on the periphery of the diagram, and shows the decision nodes where one or another factor, or several, will impel toward ongoing disease or to a different outcome.

Males are more likely to become carriers than are females, and male carriers are more likely to remain carriers and develop chronic liver disease and cancer. There appears to be a genetic predisposition to the carrier state, and the development of hepatocellular carcinoma. This may be related to the genetic control of the human immune system, to the control of other factors, or to as-yet undiscovered genes that directly affect the probability of chronicity.

Environmental factors include elevated total body iron stores, which increases the probability of cancer in those who are chronically infected with HBV. Iron store levels depend in part on dietary iron intake. Elevated levels of arsenic in potable water also appear to increase the probability of HCC.

Tobacco and alcohol use, which are behavioral characteristics, also appear to affect the course of the disease. HBV interacts with other viruses, including the other hepatitis viruses, with HIV, and with malarial infection; all of these can affect the course of the infection.

The detailed interaction of exogenous agents and genes can be illustrated by the impact of aflatoxin, a potent carcinogen elaborated by species of the fungus *Aspergilus,* which can infect improperly stored grains and other foodstuffs. In a study in China (Table 1), it was shown that exposure to aflatoxin and chronic infection with HBV has a multiplicative effect on the risk for HCC. Individuals neither infected with HBV nor exposed to aflatoxin were assigned a risk ratio of 1. Those who had increased levels of aflatoxin but were not infected with HBV had a significantly increased risk, as did those infected with HBV but without increased aflatoxin levels. However, those exposed to *both* agents had a 60-fold increase in their risk!

Even with this increased risk, there were some people who did not succumb. Additional information is now available that can explain this, in part. McGlynn and her colleagues reported the effect of genes segregating at two different human gene loci that produced protein enzymes that detoxify aflatoxin. Each of these gene systems are polymorphic, and an individual, depending on the genes he or she has inherited, will or will not effectively detoxify aflatoxin. Exposure to aflatoxin also appears to be associated with a mutation in the p53 system that can affect carcinogenesis. Hence, the individuals who have the alleles that produce the less effective detoxifying enzymes will be more likely to have the mutation in p53 and DNA damage, and will be at increased risk for HCC (Figure 3).

A great deal is known about genetic and environmental factors and their interactions that lead to cancer of the liver. Knowledge of the genetics has helped immeasurably to understand these phenomenon, but intervention for prevention, and possibly treatment as well is based not on changing the host's genome, but in intervening to protect against environmental factors. A large percentage of HCC can be prevented by infant vaccination against HBV, and this is now in common use. The contamination of food by *Aspergilus* can be prevented by appropriate storage methods; iron storage can be controlled by avoiding supplemental iron and other measure, and the levels of arsenic in water can also be controlled. It is likely that many other cancers and other diseases will yield to similar practicable methods.

A disease-free life is a heady promise, but it is already a blessing that many enjoy, and one can hope that by continuing research, it can in the future be extended to many others.

Table 1. Risk Ratios (95% CI) for Hepatocellular Carcinoma Associated with and without Chronic Infection with Hepatitis B Virus (HBsAg) and with Urinary Levels of Aflatoxin in a Population from Shanghai

HBsAg	Aflatoxin	
	Negative	Positive
Negative	1.0	3.4 (1.1, 10.0)
Positive	7.3 (2.2, 24.4)	59.4 (16.6, 212.0)

SOURCE: Adapted from Qian et al., *Cancer Epidemiol. Biomarkers Prevent.,* 3:3–10, 1994.

SUSCEPTIBILITY OF HEPATOCELLULAR CARCINOMA IS ASSOCIATED WITH GENETIC VARIATION IN THE ENZYMATIC DETOXIFICATION OF AFLATOXIN
K.A. MCGLYNN ET AL, 1994

1. MUTATION IN p53 TUMOUR SUPPRESSOR GENE IDENTIFIED IN HCC IN SOME POPULATIONS.

2. MUTATIONAL HOTSPOT IN CODON 249 OF EXON 7 POSTULATED TO BE RELATED TO AFLATOXIN B1 (AFB1) EXPOSURE.

3. MICROSOMAL EPOXIDE HYDROLASE (EPHX) AND A MEMBER OF THE GLUTATHIONE-S-TRANSFERASE MU FAMILY (GSTM1) ARE INVOLVED IN AFB1 DETOXIFICATION IN HEPATOCYTES. THESE LOCI ARE POLYMORPHIC.

4. EPHX MUTANT ALLELES (EM); AND GSTM1 NULKL ALLELE (GN) OCCUR IN MANY POPULATIONS.

5. EM AND GN INCREASED IN PERSONS WITH AFB1 ADDUCTS.

6. EM AND GN INCREASED IN PATIENTS WITH HCC.

7. EM AND GN INCREASED IN PERSONS WITH P53 MUTATION.

CONCLUSIONS:
Persons exposed to AFB1 who have an EPHX mutant allele or who are null at the GSTM1 locus may be at increased risk of developing,

1. AFB1 induced DNA damage.   2. p53 codon 249 mutations.

3. HCC.

These associations are independent of HBsAg status.

Figure 3. Features of the relation between hepatocellular carcinoma and enzymatic detoxification of aflatoxin. SOURCE: Adapted from McGlynn et al., *Proc. Natl. Acad. Sci. USA,* 92, 2384–7, 1995.

New Knowledge for Health: Response

Michael M. E. Johns, M.D.
Vice President for Medical Affairs, Professor of Otolaryngology–Head and Neck Surgery, and Dean of the Faculty, Johns Hopkins University School of Medicine

I thought that I would start with a summary, but in that summary I am going to raise more questions than give answers. Perhaps my message is quite simply that we as professionals must find the answers to these and to other questions that science will raise for us as members of the global community.

In his chapter Dr. Blumberg has covered a number of areas and has suggested important issues that undoubtedly will preoccupy us for many years to come. Perhaps what stands out most in this discussion is the extraordinary new levels of knowledge to which we shall soon have access and the extraordinary levels of wisdom to which we must aspire in order to make the most of what we will know.

We will attain new levels of understanding because we will be generating unprecedented amounts of knowledge and because we will have enormously expanded computational data capabilities. Any scientist contributing to the genome database will collect billions of bits and bytes of information about the molecular basis of the genome. New advances in computer science and computational biology will provide us with the capacity to uncover links, patterns, and associations that we never would have suspected.

For instance, we are just beginning to see the possibility of whole new areas of knowledge based on new discoveries of evolutionary and functional relationships. An example is the discovery that vision and hormone action involve similar signal transduction. This is a tremendously powerful insight. In short, it is difficult to overestimate how much we will know about our genome in 10–20 years and the impact that this knowledge will have on our capacities to cure and prevent disease.

No doubt, much more biology will be done on the computer. Computational biology, structural biology, and combinatorial analysis are developing powerful new software and mathematical techniques that can determine the structure of drugs and mimic actual molecular interactions. All of these will grow in importance.

Also of increasing importance will be the need for a larger cadre of clinical investigators who can sort out the efficacy of hundreds of new drugs that will be developed with increasing speed using these new capabilities.

What is easy to underestimate is the impact that this knowledge will have on the organization of medicine, on our roles as physicians, and on the ways in which we allocate health care resources in our society. Knowledge, after all, is power. Biologically at least, we are going to have a lot of both new knowledge and new power.

One very important question is, "Who is going to control all of this knowledge and the power to employ it?" Right now, there is a growing trend to have much of the traditional physician and patient control over the employment of medical knowledge restrained by certain types of delivery organizations and financial considerations. This is just an initial skirmish. What may prove to be a much bigger battle is the control of knowledge produced by biomedical science and medicine over the next half-century.

First, following Dr. Blumberg's lead, let us look at a few ways in which our knowledge will affect us. Curative gene therapy aimed at somatic cells in general probably will have only a modest impact on medicine and on the organization of health care delivery as we know it. However, it will raise new issues about societal costs and the ethics of rationing care. Therapies aimed at germinal cells will be far more controversial and have far larger implications for the organization of medicine, societal values, and health care costs. It is easier to see how the former therapies will play out than it is to see how the latter will.

At present, we know how to make decisions and choices about treatments for individuals—at least we think we do. In fact, that is what we do best. On a case-by-case basis, patients and physicians still basically make treatment decisions, even if that has been made more difficult by the types of third-party interventions and new cost-based considerations that recently have challenged and changed the calculus.

To the extent that genetic interventions begin to occur at the somatic cell level with individuals, the basic physician–patient relationship should not be altered significantly. However, the ability to intervene at the molecular level in somatic cell disease processes likely will generate considerable controversy at the socioeconomic and ethical levels.

For instance, let us assume that we arrive at a point where we have developed our capacities in what can be called molecular epidemiology and can predict susceptibility to lung cancer in each of the individuals in a population (e.g., of the seventh grade at your local middle school) with a fair degree of certainty. We will know the environmental risk factors, with cigarette smoking perhaps the most important of many. If this knowledge is then transmitted on a systematic basis to these students so that each is made fully aware of the risk factors over a period of years, what are the ethical or economic implications for individuals who disregard the known risk and at age 50, after almost 40 years of smoking, are diagnosed with lung cancer?

The first obvious—or at least likely—private insurance industry response might be to link insurance premiums to individual risk factors and personal compliance. If you know the risks and the risks are high, and if you take the risks and smoke, for example, then your premiums are high. If you comply with a low-risk regimen where you do not smoke, your premiums are lower. It is certainly possible to foresee an insurance industry assembling a wide variety of such risk factor profiles and crafting premiums to reflect genetic factors and environmental compliance. This would be the ultimate in risk-adjustment factors.

This new genetic knowledge is going to raise many issues about treatment decisions, the allocation and distribution of health information, privacy, and much more. What if society's inclination were to link health care spending in some way to personal compliance? We already have this argument going on between cigarette manufacturers and those who claim injury from these products.

Suppose these middle-schoolers reach age 50, are poor or disabled, and therefore are insured through public programs. This, of course, assumes that such programs still exist at that point. What if a high public priority is to limit health care spending and one important part of that policy in this new age of molecular medicine is to hold individuals more accountable for certain behaviors that substantially increase health costs? What if public policy wants to preclude the lung replacement surgery that is available, although at substantial cost, and to limit total spending on the lung conditions of those with relevant risk factors who did smoke or who work in an environment identified as high risk?

One can imagine any number of scenarios in which this type of issue could arise. What is more difficult to imagine are satisfactory solutions. The implications of interventions in germinal cells are even more far-reaching.

So let's move on to some implications beyond all of this. Most important in my mind is the need to realize that our new knowledge will bring with it both new power and new responsibilities. Not only we, but also the public, will have much more information and many more choices—often very difficult and expensive choices. Some in medicine may think that the new issues, especially the ethical, economic, and social ones, are not the concern of biomedical scientists and physicians. You might say that we developed the science, and someone else must figure out its broader implications.

My main concern with the future of knowledge in health care is that we will not cede "jurisdiction," to use an important concept from the legal lexicon, for any of these issues. This forum illustrates the need to understand these issues as part of the evolving and growing purview of the medical profession. Knowledge is power. As medicine becomes more powerful, our professionalism must become more developed.

Actually, I prefer to say that as our knowledge and our capabilities become more powerful, our professionalism must become more profound. That is one reason why at Johns Hopkins we have introduced a four-year required course of study about the role of physicians in society, in which our students study the implications of new knowledge and evolving technologies. Future generations of physicians and scientists are going to have a more profound understanding of the knowledge and capabilities they possess and an ability to play a leading role in defining and controlling the application of this knowledge. Otherwise, medicine will be reduced to the level of a craft, physicians will lose their status as professionals, and society will lose a great deal more.

I have jotted down a few of the implications of this new knowledge for our academic health centers, and they are all in the form of questions relating to this new science and the potential that we see coming from it and its applications. Fundamentally, we need to ask ourselves what kind of students we will select. What kind of faculty will we recruit for this new era? How do we organize for the future in terms of departmental structure? Some of that has been discussed for the clinical side. We must ask the same questions about the scientific side. What kind of facilities do we need? Will we need any? Will there be virtual universities, and will everybody work at home? How do we integrate where we are going into the larger university? How does this new information become part of our daily lives? Will anybody be there in line to pay for it?

There are many strong forces acting in the medical profession at the moment that would happily see us become more like employees and line workers, technicians, or skilled crafts people. The nationwide skirmish going on right now about who makes the call in the application of health care knowledge may presage larger battles to come. There is only one way we can address this, and that is not individually, but as a profession.

The Workforce for Health

Edward H. O'Neil, Ph.D., M.P.A.
*Associate Professor of Family and Community Medicine, and
Codirector, Center for the Health Professions,
University of California, San Francisco; Executive Director,
Pew Health Professions Commission*

I will limit my remarks to two things. The first is to give you some sense of the challenges to the health profession, which have already been addressed by others. Then I would like to focus by looking at specific health professions and offer some comments on the academic health center and its future.

Figure 1 is a picture of the health care system in the United States, circa 1981. Essentially, I would like you to focus on the reality that all of the rules we developed about health care over 50 years—who could practice, what they could practice, where they could practice, how much they were reimbursed, who provided oversight, and who did the policing—were all controlled by the professions. There was token, if any, public input into the design of that system. It emerged in the 50 years after the Second World War principally because in 1950 this country had 7 percent of the world's population and sold 49 percent of the world's goods and services. We could afford a system that was structured in that way.

Now, what is emerging clearly—whatever it may be called—is a system of integrated delivery, whether it is managed, profit driven, or not-for-profit. That system will have essentially three values: One is maintaining or lowering the cost of care, which is essentially the only value that is operative right now and will be for the next four or five years. The second is increasing patient satisfaction as a consumer. We can tell ourselves all we want about the sanctity of the patient–physician relationship, but the reality is that patients have not perceived this as something that has been responsive to their needs and have felt unempowered in that environment. Although the relationship between the provider and the patient is considered the most sacred relationship in health, I suspect that the most important relationship is between one-

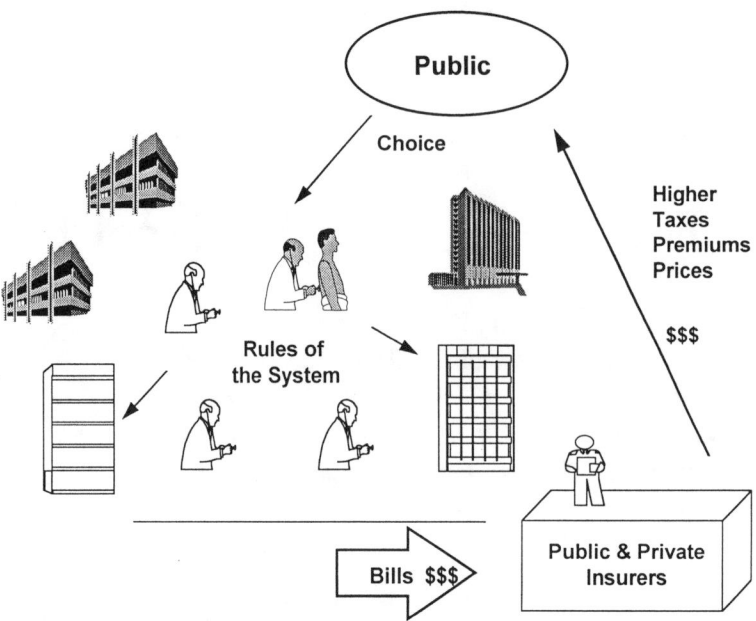

Figure 1. Health care in the United States, circa 1981.

self and one's health. The final value involves improving the quality of patient care outcomes.

The system is moving faster now than even the most aggressive prognosticators thought possible as recently as two or three years ago. By the end of this decade, I believe that in most major markets of this country, 90–95 percent of the population will be enrolled in some sort of integrated system, not owned necessarily all in a single place, but integrated in the sense of long-term contracts or exclusive contracts. Already, in northern California; Portland, Oregon; Seattle, Washington; and the Twin Cities, such patterns of organization are emerging.

The nature of the integration is almost unimportant; horizontal or vertical integration is almost unimportant. However, the size of the system is critical. How the systems come together is something for consultants and attorneys to work on, not something that will have anything ultimately to do with the health of the nation.

This integration is occurring in three phases. Let me focus on one dimension of each, because in each of the three phases there is a cost-control and system value-added dimension that presses severely on health professionals and their work.

The first stage is assembly—the cobbling together of systems. That is going on right now. To try to make sense out of this and then project into the future is foolish because, next year, someone will buy the system on which that analysis is based and the analysis will be useless. In this first phase, limiting access and reducing fees are essentially the means of controlling costs. This is what is pressing most on professionals as they take on the responsibility for that risk.

The second phase sounds terrific—the integration of this far-flung set of health care resources in management information and decision support systems. This is the promise, the upside.

The downside in integration is having to remove excess capacity from the system. In independent hospitals it made sense to have an operating census of 20 percent. In putting together a system of four or five hospitals, it makes no sense to have each operating with that kind of enrollment. So that is where we remove perhaps 30–50 percent of the hospitals; maybe 40–60 percent of the hospital beds; from 100,000 to 150,000 of the physicians, all of them specialists; and perhaps 200,000–300,000 hospital-based nurses—not because we change the intensity of care delivered in the hospital, but because the hospital is not operating any longer—the work force simply is not there.

Finally, we arrive at a management phase. If there is a silver lining in this rather brutal process, it is in management. If there is an American genius in what we are about, it is bringing a brand new approach to the organization of health care. This not only will redesign the health production function to yield the higher value that has been discussed, but will also change the process by which we produce care. Determine what the inputs are, professional and other, that will actually produce a particular outcome, and then radically redesign in that context the practice of health professionals.

Why the professions are key to this change is obvious, but let me mention a few reasons. There are 10.5 million health care workers in the country, and we spend about $20 billion–$25 billion on education across the continuum of health professions. Health care itself is a labor-intensive process: 70–80 percent of the expenditures in any operating unit will be for labor costs. The professions control cost, waste, and innovation by how they work.

The existing professional structures are not amenable to change. We have essentially a set of 19th-century work rules operating in a 21st-century institution. We have this system that has been dominated by supply. What is it that we want to provide? What is best for us? How do we want to array the community needs to increase demand?

If you do not think we have had a bias toward the supply side in responding to the public's problems, why is it that we build neonatology units in response to low birth weight and prematurity?

All of the professions are tied to this aging model. Medicine is overly focused on individual practitioners providing clinical service to individual patients who present with acute care treatment needs.

Nursing is oriented to an undifferentiated practice across most ranges of nursing today, particularly within a hospital setting. By continuing to operate in this way, the emerging system is demanding different roles for nurses.

Pharmacy is still focused almost exclusively on the physical delivery of a drug, even though we know that the guarantees of quality and other dimensions of value added by the pharmacist can be provided in many other settings.

In allied health, the more than 200 different allied health professions still define themselves by the technology or the therapy that created them and build walls around those technologies and therapies to protect themselves.

The transition that is likely to occur over the next decade is one from protectionism, to use a very stark and harsh word, to pragmatism for the professions: from professions defined by a scope of practice—I might even say captured by and suffering from a scope of practice—to a health system that is redesigning itself to meet new values. These values involve particularly, a capacity to lower costs, to deliver health quality at a higher outcome, and to enhance patient satisfaction. This may or may not have anything to do with the scope of practice. The health system itself increasingly has the power to make those decisions and looks to the professions to assist it.

Professions have been unwilling or unable to capture or control quality. There is nothing in 40 years of the literature to indicate that the professions have been up to that task, unless the violation of quality was so egregious as to involve the criminal justice system. Then and only then do the professions have a record of actually policing their own in any effective way.

So now we have quality standards not only developed from the system but demanded by the purchasers of the system. That power shift, more than anything else, will drive and create the new reality.

Controlling information and access to that information has been central since the Middle Ages to the definition of what a profession is about, but we now know that information is abundant, cheap.

The first step is the redesign of the health workplace. This is going on right now with little, if any, public discussion. The market-driven changes are pushing faster and deeper into reform than anything proposed by the Clinton administration.

The next step is reregulation of professional practice, not its deregulation. We do not need to leave the professions or the public out in the cold against these terrible mechanisms of the market—the profit-driven managed care systems. What we need is a set of regulations that serve the public interest at the end of this century, not the end of the last century. Then must come right-sizing—the politically correct word. There is some upsizing necessary in certain professions, but for a couple of important professions, there must be some downsizing. Finally, there is the restructuring of education.

Redesign of the workplace focuses on outcomes, quality, cost, and the information base. It is performance driven. We have touched on all of these.

Flexible, innovative, and adaptable are not words that have been hallmarks of the health professions. We must recognize those professions that once were viable but are no longer useful, and we need to get on with the process of team-oriented education and training. Then costs can be determined.

Reregulate professional practice. Practice acts must be based on demonstrated initial competence, and professional boards must focus on changing public and care system needs. The role of the consumer in all of this must be enlarged. Continuing competency requirements must be required as a part of that. The entire process must be focused on quality performance, or it simply will not be worth the time invested in it.

Right-size the professions or reduce the number of physicians produced. Too many physicians are being produced by both the graduate and the undergraduate medical education systems. This number cannot be sustained in the system as it is now, much less as it will become.

Reduce and redirect nursing practice toward four-year and advanced practice programs. The size of nursing programs at the diploma and associate degree level must be reduced.

Reduce the number of pharmacists produced. In a system that will rapidly take advantage of information, communication, and transportation technology to distribute drugs, we simply do not need pharmacists handling drugs unless you like the opportunity to visit your local drugstore.

Expand public health programs and increase the number of public health professionals. Several others have said that this is a great day for public health. In the evolution of the management phase, we eventually move to management of the health of populations. We do not have them integrated enough into clinical disciplines. The number of multiskilled allied health workers must increase—not the number of technically driven allied health workers, but the number of allied health workers who are multiskilled.

Make education accountable for cost, time, and performance. If you liked managed health care, you are going to love managed education, because it will be in terms of performance and accountability standards. New skills must be created and offered for professional education.

Remove time and place boundaries. It makes no sense to have a resource tied up in an institution. Move it to an ambulatory community setting. Clearly, we have made major strides in that direction.

Merge with integrated care systems. It makes no sense to affirm a department of community and family medicine today but build it like a department of medicine in 1970. That was the wrong image: research dollars were abundant, and patients were coming into the hospital. To build a primary care department today, build it at Kaiser Permanente; then figure out how to appoint the faculty back to your academic health center. Then create a funding-based educational mention not to research, not to service, not to indirect recovery.

To speak specifically about a few professions, physicians are oversupplied. Figure 2 is a picture of one estimate of specialist positions from the Pew Health Professions Commission. If we continue in this way until 2020, there will be perhaps 190,000 too many specialists for that environment. The horizontal bars represent high and low projections from John Weiner for managed care enrollment.

The challenge for medicine is to build professional identity around managing risk and adding value, not around a set of practices, not around a scope

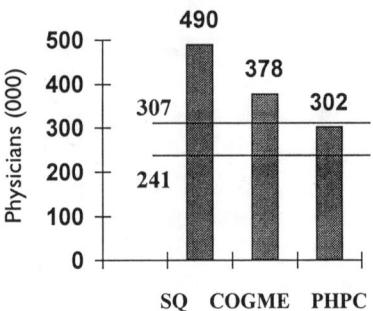

- Oversupply of physicians
- Demands of system/ purchasers
- Consolidation of plans
- Efficacy of integrated group practice
- New demands on professionals

Figure 2. Forces reshaping physician practice.

of practice, and not around dominance within the system. What can physicians do whose quality cannot be matched by other professions?

Enhance relationship-based care. I believe that the individual ultimately will be looked upon as the place where we add value to health care ourselves. The physician–patient relationship is important, but it needs to be recaptured because frankly it is somewhat tarnished.

Reorient on a primary care paradigm. One obvious feature of this emerging managed care system is that it will be integrated through the primary care disciplines, not through specialty disciplines. This is not to denigrate specialty disciplines; they are still critically important for what they do. They just seem to have proven to be a very expensive way to integrate the entire health care system.

Create new relationships between primary care and specialty care. The relationship must be basically different. One of the most important challenges to medicine is to create a single pathway to a generalist's career so that, rather than having general internal medicine, family medicine, and pediatrics, we have a general medicine pathway with pediatric, geriatric, adult male, and adult female opportunities for specialization.

Ensure continued clinical innovation in the subspecialties. We must recognize this as an important dimension of the health care system and of the biomedical research system, and figure out a way to maintain it without orienting the entire care delivery system toward that model.

Downsize the profession, which I have already mentioned. Use information and communications technology to inform and shape an advanced practice. If there is one profession that will make the most of this opportunity it is medicine. This image of having all information inside the head or in the hands of an individual practitioner is an image from a bygone day. Medicine must embrace this new technology to bring the values that it possesses to health care.

Most nursing still takes place in hospital settings. Figure 3 shows the ratio of registered nurses (RNs) to beds. The average length of stay in days and cities that decrease in the intensity of managed care are also shown. This is a

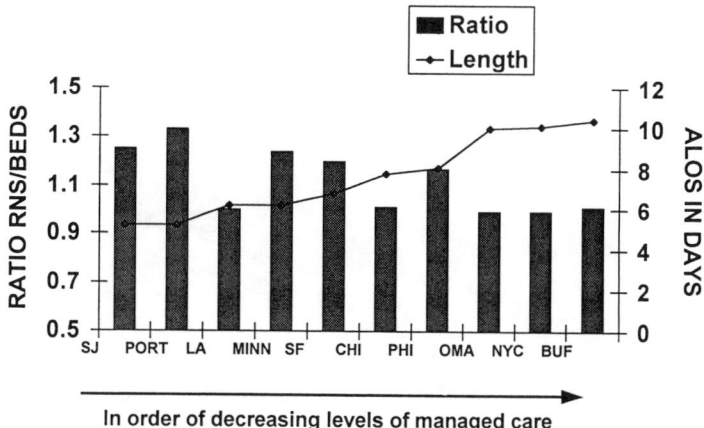

Figure 3. Average length of stay (ALOS) and ratio of registered nurses (RNs) to operating beds, 1992. SOURCES: Statistical Abstract of the United States, 1987, Bureau of the Census, page 90; The Registered Nurse Population, Findings of the National Sample Survey of Registered Nurses, March 1992, Bureau of Health Personnel, Department of Health and Human Services; and The AHA Profile of U.S. Hospitals, 1993/94, American Hospital Association.

Group Health Association of America measure of the intensity of managed care. You can see the correlation between the intensity of managed care and the ratio of RNs to beds. So if a shorter length of stay is a by-product of more intensive managed care, nurses are being utilized in a much higher ratio to beds than in a more conventional setting.

The other point illustrated by this graph is that the complexity of care—the case mix that the nurse sees in the hospital setting—has changed, and the skill base demanded of a nurse is much more sophisticated than in the past.

So the challenges for nursing are to reduce the training capacity radically. The Pew Health Professions Commission recently recommended a 20 percent reduction in training programs for nursing. This is an inadequate beginning for a decrease in the size of training programs. For those who remember the cyclical boom and bust in nursing, remember that boom and bust represents a labor market phenomenon driven on a health care system that was growing from 1960—6 percent of the gross domestic product (GDP)—to this year—15 percent of GDP—with cyclical adjustments. We are talking now about what happens when you reduce insurance premiums by 12 percent over two years as we have in California. You begin to close hospitals, which is a structural adjustment, not a long-term cyclical labor market adjustment.

Expand advance practice training facilities and programs for nursing. Simplify the training degree and practice structure.

One thing we have heard over the past few years is that nursing is its own worst enemy, with a proliferation of degrees, titles, and pathways.

- Clinical income represents major source of support
- Directly or indirectly subsidized education and research
- Most of it going away or reduced margins

Income Sources, 1992

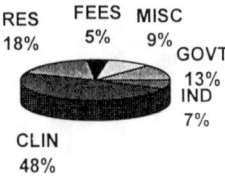

RES 18% FEES 5% MISC 9% GOVT 13% IND 7% CLIN 48%

Figure 4. Vulnerable health centers.

Delineate degree practice dimensions for all levels of nursing and focus on care management.

We have heard a lot about the vulnerability of the academic health center. The important message of Figure 4 is that historically, about half of our income has come from clinical services in the last 10 years. We have built an enormous infrastructure.

Figure 5 shows what is happening in research—a familiar story to all of you. The bottom line represents RO1s awarded by NIH; the top represents those reviewed. Not only will we produce more specialists, we will also have a similar growth curve for the production of biomedical investigators. They cannot all have five graduate students and three postdocs in their labs and expect to be funded on a fairly stable base.

In the academic health center this is the challenge. Reduce the demand for specialty education. With declining federal and state support, it simply is not coming back in the way we expected. There needs to be more focus on primary care. Many of our academic health centers have an inadequate patient base for teaching. The patients simply are not there any longer.

More focused demands are being made of health for the health care system. The demand for specialty services, especially middle-tier services—not organ transplants, but undifferentiated cardiology, gastroenterology, and radiology services—is declining radically. Capacity is inadequate to support hospital settings and services. The opportunity exists to develop primary care networks, and the loss of income stream is perilous to the continued existence of hospitals.

On the organizational level, academic health centers will lose considerable parts of their revenue streams, not just from patient care but from research dollars and other sources. We must develop new partnerships. That has been addressed earlier and is absolutely critical for the future. We must incorporate information technology. Hospitals simply will not look like the citadels of the past.

There must be a greater range of diversity of institutions. We have to separate the teaching, research, and clinical service missions and be willing to pull them apart. I don't think we have good figures for what it costs to hold them together, but must find ways to pull them apart and then use technology to relate them as they need to be related.

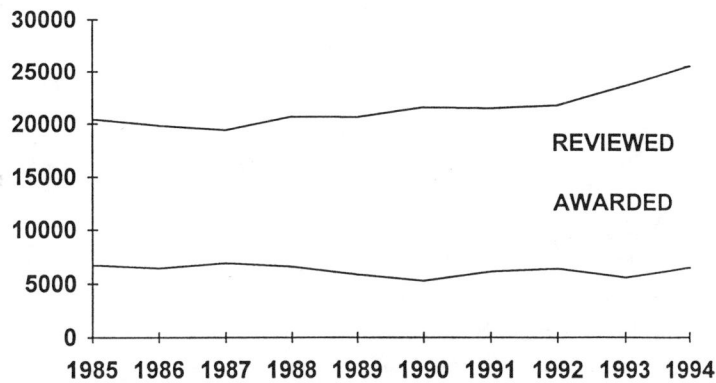

Figure 5. Research changes: competing research projects, 1985–1994. SOURCE: NIH Extramural Trends FY 84–93, U.S. Department of Health and Human Services, October 1994, page 37.

Conduct clinical service only as it is needed by education and research. Right now we are a proven high-cost provider. We need to develop partnerships with the care delivery system that permit us financially and in terms of patient flow to carry out the education and research mission.

Integrate education and research with other parts of the care system. These are important commodities. We need to make more of our research capacity—our ability to develop and disseminate new knowledge—because knowledge is going to be the coin of the realm in the new system.

Create a virtual academic health center, not one located on a single campus, but one in which research might be spun off; patient care services might be carried out in efficient, patient-oriented care delivery systems; and the educational mission might be carried to the research lab, the patient care delivery site, or the lecture hall. Then create a stable and sustainable financial base.

Finally, Kierkegaard said that "life can only be understood backwards; but it must be lived forwards." Oftentimes with these kinds of projections, we try to give the perfect argument for the future. That is absolutely impossible. However, it is important to be actively engaged in the next 10 years so that we can accurately reflect and understand where we have been.

The Workforce for Health: Response

Paul F. Griner, M.D.
*Vice President and Director, Center for the Assessment
and Management of Change in Academic Medicine,
Association of American Medical Colleges*

D r. O'Neil and other members of the Pew Health Professions Commission should be acknowledged for their important contributions. They have focused our attention on a number of key factors that should influence planning of the health professions work force for the 21st century. Among these are the health status of the population, the growth of knowledge of disease and of the technology to apply that knowledge, and the paradigm shifts that medical centers will have to embrace to help them achieve their missions.

HEALTH STATUS

On some of these points, the Pew Commission report does not go far enough. On others, the challenge is not what but how to address the issue. Perhaps we should start by focusing on the ultimate goal, that is, improving the health status of the population. One of the key recommendations made by the Pew Commission is for a work force that is more oriented to improving the health of the entire population and more inclusive in how it defines health. I agree but suggest that the implications for the number and mix of providers in the health care field are *even more profound* than has been suggested. The health status of the public will continue to be determined as much or more by social, economic, and cultural determinants than by the amount and quality of health services as we now know them. I emphasize as we *now know them.* Morbidity and mortality from physical and mental abuse, various addictions, and other preventable conditions, compounded by poverty, lack of education, and the pace of change, will continue to exceed illness and death due to impaired health from nonpreventable causes. This assumption has implications both for the boundaries of medicine and nursing and for the desirable mix of professionals best suited to address these challenges. It is no longer appropriate to take the position that these are factors outside our realm. While the health professions cannot be held accountable for impaired health status attributable to poverty, crime, and the like, the professions must expand their definition of the boundaries of health care and find ways to interact better with their communities—more involvement with schools, with neighborhood groups, with individual families. The desirable mix of professionals best suited to address these challenges is a point for discussion. Social workers, psychologists, teachers, and parent role models may play more important parts than physicians and nurses. In any case, the implications for both the nature of the work of health care and the mix of team members best suited to that work are significant.

KNOWLEDGE AND TECHNOLOGY

The growth of knowledge of disease and of the technology with which to apply that knowledge is another important variable that has implications for the work force. Change in medical practice is being driven as much by these factors as by the way health care is being delivered and paid for. I suggest that there are more uncertainties around these determinants than around the work force needed to function in a reorganized delivery system. Let's examine the implications of the knowledge being accumulated concerning genetically determined or mediated disease. Will this knowledge influence the number and mix of providers? If so, in what direction? In theory, a reduced burden of illness would be expected from the application of this knowledge. On the other hand, those at greatest risk for disability and death from the more prevalent diseases may have even less access to diagnostic or therapeutic technology than they have today if entitlement programs are scaled back too sharply. Will the lowered resource use anticipated from better knowledge of those *not at risk* for coronary heart disease be neutralized by the aging of a high-risk minority population whose access to health services is reduced? It is your call.

WORK FORCE

Notwithstanding these considerations that may influence the number and mix of health care professionals needed, it *is* important to acknowledge an oversupply of physicians and to discuss strategies that might be used to address the problem. The Pew Commission report calls for a limitation on the number of residency training positions, limits on the number of international medical graduates receiving training in this country, and a reduction in the number of U.S. medical schools by as much as 20 percent. There are at least two important points to be made with respect to this recommendation. First, it is premature and potentially damaging to the interests of Americans who seek training for careers in the health professions to call for a reduction in the

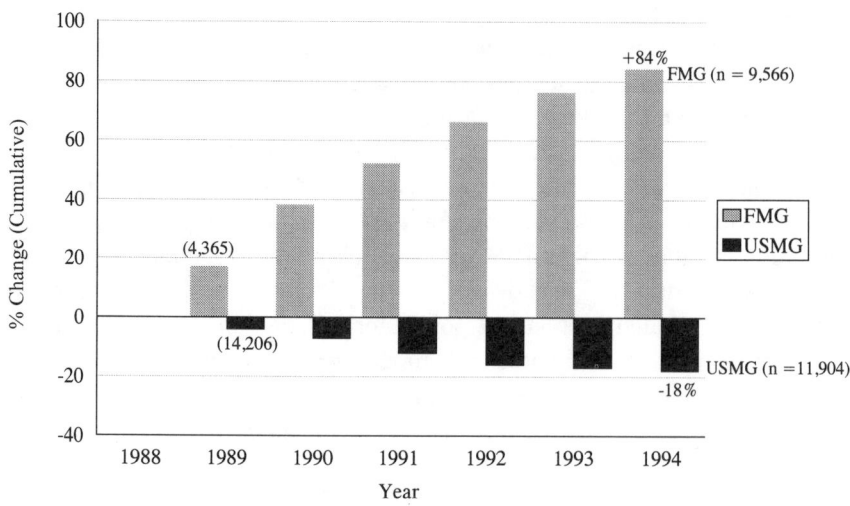

Figure 1. Cumulative percentage change in internal medicine residents, 1988–1994. FMG: foreign medical graduate; USMG: U.S. medical graduate.

number of medical schools or students *in the absence of a clear plan* to address the excessive number of residency positions in this country and to limit the number of international medical graduates who remain and practice here. The emphasis needs to be on the sequencing of these objectives. It is easier to talk about reducing the number of U.S. medical schools or students than it is to agree on strategies to address the number of residencies and the maldistribution of American and international graduates occupying these residencies.

To reinforce that point, Figure 1 shows the cumulative percentage change in the number of foreign medical graduates (FMGs) in training in internal medicine residencies in the United States since 1989, compared to the change in number of American medical school graduates (USMGs). An 84 percent increase in the number of FMGs contrasts with an 18 percent *decrease* in USMGs, the result of a continuing increase in the number of residency positions offered and a decline in the number of Americans applying. The result is that by July 1996, there will be more foreign than American medical graduates in residencies in internal medicine in the United States. This trend will be seen *across the spectrum* of residencies in the United States unless or until a strategy is developed that leads to a reduction in the total number of positions and to limits on the number of FMGs occupying them. This strategy must recognize the demand for foreign-educated physicians as a means of satisfying service needs in our teaching hospitals. Sufficient funds must be made available to those hospitals that provide disproportionate medical services to indigent patients and depend on residents who graduated from foreign schools to provide those services. Additionally, leaders of academic medicine must begin to lead by weaning their institutions away from dependence on graduates of foreign schools. This dependence relates not only to the provision of service. It exists in the biomedical research arena as well.

A second point needs to be made in response to the recommendation to close some U.S. medical schools. Medical schools do not exist in isolation. The missions of research and community service are intimately tied to that of education. For most medical schools, to close the school is to eliminate the social good that emanates from its investment in research and its contribution to the health of the community. *If* a reduction in the number of students or schools is ultimately found to be a desirable strategy, it makes more sense to some of us to talk about consolidation of medical schools and their teaching hospitals of the kind that is now being played out in New York City, Boston, and the San Francisco Bay Area.

CONCLUSION

Finally, it is worth commenting on whether regulation, the marketplace, or both, will represent the best approach to the necessary downsizing of residency training positions. Regardless of one's intuitive position, it seems clear that better and more timely trend data are required to suggest the best strategy or mix of strategies. Figure 2 shows changes in numbers of trainees in cardiovascular medicine in the United States over the past 5 years. This is the subspecialty, as you all know, that has the greatest oversupply. The inexorable growth in numbers appears to have abated in those areas of the country where oversupply is most quickly recognized, areas where the growth in managed care has been most rapid. So we see a 5 percent reduction in number of positions among academic medical centers in the West; a smaller decline, but a decline, in the Midwest; and a flattening in the South. Only the Northeast has not yet recognized the imperative. Some of us are of the opinion that the number and mix of subspecialty training positions will begin to be influenced

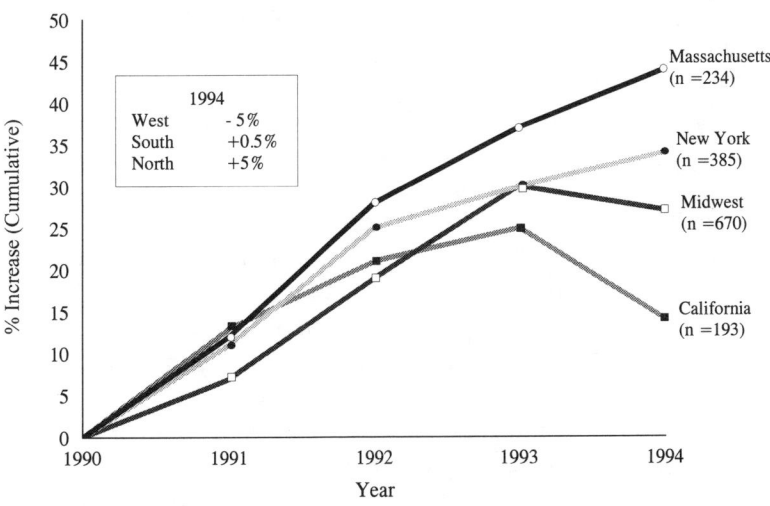

Figure 2. Cumulative percentage increase in cardiovascular medicine fellowships by state and region, 1990–1994.

sufficiently rapidly by the marketplace that regulation will not be required. This will most likely *not* be the case for the number of core residency positions for reasons that have already been mentioned.

Let me conclude by suggesting the obvious; that is, many past work force projections have missed the mark by a large margin. Although what is proposed today may seem reasonable, there remains much to be learned about the many variables, and their direction, that will continue to influence the actual demand for the health work force. We should continue to monitor and measure the impact of these variables as we continue the debate.

DISCUSSION

DR. LITTLETON: My name is Preston Littleton. I am with the American Association of Dental Schools. This is just a reminder that an academic health center by definition consists of more than a medical school and a teaching hospital. It was really only late in the conversations that we owned up that there are schools of pharmacy, nursing, and dentistry. As we look at these changes, it is imperative that we look at them in a collaborative way as to how all of our professions and our colleges are going to be responding to these changes in market forces. I just wanted to remind everyone that there are more of us out there than just the schools of medicine and the teaching hospitals.

PARTICIPANT: As long as we are on that topic, I would concur that we need to give more recognition to the role that other health professionals play. I believe that as we do a lot of our work force planning, it seems that there is sort of a nod and a wink to the role of nursing and to dentistry, pharmacy, physicians assistants, and other health professionals; but there is not enough focus on what it means for us in the production of primary care physicians.

In the spirit of this event, we also have not looked out very far on this agenda. I have heard a lot about telematics, virtual reality, and virtual surgery; but I do not believe we recognize or have stretched our imaginations far enough to understand how many primary care physicians there will be in the year 2020. If our epidemiologists could tell me the population of this country at that time, whether it will be 260 million or 270 million, I expect that we will probably have, if one would assume that we would have such a gracious society, that everybody would be computer literate, and be involved in having direct access to all of these telematics and everyone would be delivering a significant amount of your own primary care.

We need to reflect on that because it has profound implications for what we are doing in all of the health professions, not only in how we are going to teach. In fact, the total numbers that we are producing could be an enormous excess for all of us. I would just like people to start putting that into their computers and begin to think a little further than 5 or 10 years and to use that imagination of the power that could come from what Dr. Blumberg has suggested in terms of our knowledge base, what Dr. Detmer has suggested in terms of what truly could be expert systems and, quite frankly, discoveries that have not even been thought of today.

DR. RUBIN: Bob Rubin. I would like to ask Dr. O'Neil, in light of his Kierkegaard quote, whether we have gone past the paradigm of the primary care physician or the primary case manager? Indeed, if 10 percent of the people account for 70 percent of the health care costs, isn't what we really need—particularly given recent articles in the *New England Journal of Medicine* and *JAMA* that show that perhaps primary care physicians are not the best managers of chronic disease like heart disease, cancer, and so on—a specialty-dominated system that uses some of the multiskilled allied health professionals that you talked about? Isn't that what we are looking for in 2020 rather than sort of this transitional mode of primary care physicians?

DR. O'NEIL: We certainly had specialty domination in the past that has not seemed to produce the kinds of health outcomes or the kind of health system that you spoke of. Yet I do believe that there is terrific promise for not divorcing specialization and specialized knowledge from the system but creating different ways to get it incorporated, as opposed to the individual practice of specialty physicians. Having primary care physicians and also having those primary care physicians relate to a specialty information base very differently than they relate to them today is as much a viable option for the future to produce the same type of outcome.

PARTICIPANT: What I would like the audience to contemplate also is, as we talk about downsizing the health care industry, academic health centers, and training programs, think about the economic impact that that will have on local communities. We should try to not look like the steel industry. There should be enough intellectual power here to think about what it means when in the Chicago or New York City begin to downsize that whole industry—downsize it appropriately and efficiently. Yet then what should people be training for as an alternative? How can we prepare for that?

DR. LARSON: Dr. Larson, from Georgetown University School of Nursing. I agree, Dr. O'Neil, with your comment on the need to downsize nursing as well as medicine, particularly the diploma and associate degree programs. However, like several other things we have talked about today, public policy does not match the direction in which we need to go. About 95 percent of Medicare funding for nursing education goes to hospitals with diploma programs, and those programs are totally antiquated.

25th Anniversary Keynote Address

Honorable Donna E. Shalala, Ph.D.
Secretary, U.S. Department of Health and Human Services

L et me begin by reading a letter that the White House just sent over to me.

Greetings and congratulations to all of those gathered to celebrate the Institute of Medicine's 25 years of devoted public service. Since 1970, the Institute of Medicine has contributed thoughtful and wise health policy analyses covering an extraordinary range of issues from mental health to Medicare, from nutrition to new vaccines, that are of great interest and importance to us all.

As we continue to work towards a high-quality, fully accessible health care system for all of our people, your work is essential to maintaining the finest possible care. I am pleased to commend the Institute for its efforts day in and day out to improve the health of the American people. Best wishes for every success in the years to come. [signed] Bill Clinton.

I want to thank the Institute of Medicine—not only for inviting me here for this special occasion, but also for lending our administration a team of truly remarkable health care leaders—from Phil Lee to Harold Varmus, from Shirley Chater to David Kessler, from Ruth Kirschstein to Claude Lenfant, and from Bruce Vladeck to Jo Ivey Boufford.

In 1971, one of your charter members, Dr. Irvine Page, challenged this Institute to "maintain its integrity, exhibit courage in its decisions, and willingly undertake study of problems that others prefer to shun."

For 25 years, with passion and perseverance, you have met that great challenge:

• 25 years of calling the health care problems facing our most vulnerable citizens by their true name: America's problems;

- 25 years of rejecting politics and focusing our country's spotlight on critical issues like drug use, vaccines, AIDS, and breast cancer;
- 25 years of helping this country improve the way it delivers health care and educates our workforce; and
- 25 years of showing the world how sustained victories in basic science can help us win the ultimate war against some of our greatest enemies, such as stroke, cancer, and heart disease.

That has never been more important than right now, because from kitchen tables to the halls of Congress, we are engaged in a historic debate about the role of the federal government.

This debate is about much more than the size of our budget and the way we allocate our precious resources. At its heart, this debate is about our values. It's about who we are; it's about what values mean to American children; and, as we peer into the future, it's about the kind of legacy we want to leave our children.

Fifty-five years ago, when President Roosevelt dedicated the NIH campus, he summed up our national commitment to biomedical research in this way. He said: "I dedicate it to the underlying philosophy of public health. To the conservation of life. To the wise use of the vital resources of the nation. I voice for America and for the stricken world our hopes, our prayers, our faith in the power of . . . humanity."

Since World War II, our historic commitment to biomedical research has spawned a steady march of progress—from the breaking of the DNA code to the mapping of the human genome. Since World War II, through years of Republican and Democratic leadership, that hope—that vision and that commitment to science—has not wavered. It must not waver today.

Like war and peace, revolutions in science have defined the ages, transformed our lives, and altered the very course of global history. Like war and peace, investments in science guarantee our national security—and must remain a national priority.

That's why, at the Department of Health and Human Services, we recruited the most brilliant scientific minds in the country to lead our national effort—leaders like Phil Lee, Harold Varmus, Bill Paul, Rick Klausner, Francis Collins, and Zack Hall.

That's why the president created the National Science and Technology Council to give us an integrated research and development budget that focuses on key national goals.

That's why, under the remarkable leadership of David Kessler, we've slashed drug approval times by 50 percent since 1987.

That's why we've supported the new science education standards from day one—so that all Americans are introduced to science from the time they start crawling.

Finally, that's why, in a time of zero growth across government, our administration has fought—and will continue to fight—for steady increases in research.

These efforts have paid off—time and time again. In just the last few years, we saw a team of researchers discover BRCA1, a gene linked to hereditary breast cancer. We helped discover the first drug treatment for severe cases of sickle cell anemia. We sponsored a clinical trial that demonstrated that AZT (azidothymidine) can reduce the risk of perinatal transmission of AIDS by 67 percent. In addition, recently, researchers demonstrated the first treatment ever for the most common type of stroke.

Have we made a serious commitment to science and research? Yes, we have. Yet, as we approach the 21st century, is it enough? No, it isn't. We are not even close.

In the classic novel, *Alice in Wonderland,* the Queen of Hearts gives some sage advice to Alice. The Queen explains that every morning, before she eats breakfast, she makes herself believe in the impossible. In fact, she makes herself believe six impossible things.

Taking my cue from Lewis Carroll, I believe there are six impossible things that must become *possible*—that we must make *possible*—to ensure that America continues to lead the world in research into the next century and beyond.

First, let's ensure that the culture of research not only survives—but thrives. We know that research cannot survive with the uncertainty of seven-year budgets and the certainty of evaporating discretionary funding.

To create real security for research itself, as well as for our scientists and their students, we must find new ways—more stable ways—of financing research in the future. We all know why this must be our top priority. Because it is only with real security that we can nourish the seeds of research. It is only with real security that we can create an atmosphere in which young investigators are pulled into science, inspired to stay there, and ultimately train the next generation of researchers.

We have built a magnificent scientific infrastructure, but it is extremely fragile. Like many of our most cherished treasures, it is hard to build and easy to tear down.

One caution—as we search for financial stability and security, we must ensure that we are making every dollar count:

• That's why, at NIH, we are conducting top-down reviews of our intramural research, so we can spend our money wisely and effectively.

• That's why we are fine-tuning the peer-review process to ensure that it continues to serve us well in funding the best science.

• That's why we must focus on prevention. The fact is, about 50 percent of all deaths have their roots in personal behavior. We must invest in all the science that holds the potential to prevent disease and help Americans to live healthier lives—from environmental research to occupational research to behavioral research.

Second, we must look beyond the bottom line of profits to the bottom line of progress. I don't have to tell any of you about the impact that cost-cutting is having on some of our most fundamental missions and most cherished institutions.

It's not easy or inexpensive to run an academic health center—I know, I ran one. Yet our academic health centers are the envy of the world. They are the places where we unlock our greatest scientific mysteries. They are the places where we educate and train the next generation of scientists. They are often the places where we care for our neediest citizens.

Their mandate is unique, and their goals will not survive in the marketplace. Together, we must address the complex—and potentially devastating—pressures squeezing academic health centers today.

At our department, we are forming a working group—led by Phil Lee—to do just that. We want to reach out to leaders of academic health centers and help find innovative ways to safeguard the irreplaceable—the absolutely *irreplaceable*—contributions that these great institutions make.

At the same time, we must ensure that we do not punish progress. Therefore:

• It is time to tackle the perverse economic incentives that discourage experimental studies and other attempts to expand the frontiers of knowledge.

• It is time to train a health care work force that can respond to and thrive in the next century.

• Moreover, it is time to protect and strengthen the treasure of clinical research. That's why Harold Varmus has convened a panel to build upon the IOM's recent report on patient-oriented research.

Under the leadership of Dr. David Nathan, a group of experts will ask some fundamental questions about how we can protect the critical role—the "translational role"—of clinical research. We need to know how we can reinforce the link between the laboratory of basic science and the living rooms of our citizens—and how we can ensure that the best science in the world ultimately pays off for all Americans.

That brings me to the third impossible thing: Let us ensure that our bioethics are as sophisticated as our science.

For every great scientific breakthrough—whether in genetics or in medicine—ethical questions will emerge, and those questions must be addressed carefully and immediately because

• we must not create a world in which our genetic map is used to deny jobs or health insurance;

• we must not create a world in which the worthy goal of science eclipses our fundamental sense of humanity, fairness, and values; and

• we must not create a world in which discoveries of the future widen the chasm between the haves and the have-nots—and send us backward.

Fourth, we need to take the long view of basic science. The promise of gene therapy, although awe inspiring, is still far from being realized. That's the tough wake-up call we received recently from a panel of experts, and from it, we were reminded of some important lessons.

• We were reminded that we must invest more in the foundation of our scientific universe—in the incremental gifts of basic science that will help us unleash blockbuster discoveries over time.

• We were reminded that we must do a better job of educating the public about what is really possible right now and what is not.

• We were reminded of the importance of honesty and self-criticism— the importance of being willing to change directions, even in midsentence— if it will help us reach our common goals.

Fifth, we cannot move forward by falling back. We cannot even begin to plan for the future of research unless we stop policies that will move this country and its health care infrastructure—public or private—in the wrong direction.

• When the Republicans in Congress try to take hundreds of billions of dollars out of our health care system and out of the hands of our citizens, that's not progress.

• When they try to take away the Medicaid guarantee of health coverage for 37 million vulnerable Americans—and replace it with an underfunded, ill-conceived block grant—that's not progress.

• When they ask us to tear a big loophole in the nursing home standards that the IOM helped create—and allow us to go back to the days when citizens in nursing homes went without vital protections—that's not progress.

• When they try to slash Head Start, Earned Income Tax Credits, student loans, and other engines that help move people out of poverty—and toward healthier lives—that's not progress.

If we allow ourselves to take giant steps backward, we may never catch up again.

That's why we believe we must balance the budget without breaking our historic promises to science, our citizens, and our country. That's why, in this budget climate, it is not enough for the American people to simply appreciate science. This brings me to my sixth and final challenge.

Sixth, every citizen must be an active constituent of science. Whenever science has leapfrogged over the general public's knowledge, we have been catapulted into confusion, stagnation, and even darkness. Just ask Galileo.

To succeed in the 21st century, our obligation must go beyond answering our most perplexing scientific questions. We must hop onto the information superhighway and ensure that critical information about science and health is only a keystroke away for our citizens. Most important, we must write the poetry of science in prose that the American people can understand.

We need a sophisticated electorate that has the context and intellectual discipline to absorb the great breakthroughs; an electorate that understands the historical, social, and economic urgency of investing in science; an electorate that cares deeply about science because it understands how science touches and benefits its lives; and an electorate willing to nourish and enhance our investments in good times and—at the very least—to safeguard them from the kinds of indiscriminate budget cuts we are seeing today in Washington. That—above all else—is the key to our collective future.

So, let us ask ourselves the following questions: When we are long gone and history books of our time have been written, what will they say about our contributions to the future of science and the future of our country?

At a time of great revolutions in both biology and information, did we rise to the challenge?

Did we keep our historical promises to our most vulnerable citizens?

Did we maintain our strong commitment—our international commitment—to basic science and clinical research?

Did we attract, train, and sustain new generations of brilliant scientists—women and men?

Did we give the American people the tools they needed to make the right choices with the only lives they will ever have?

Did we embrace our common vision and move forward on our common ground?

Quite simply, did we do the right thing?

Like the dying woman who plants a tree for her grandchildren to enjoy, every seed of science that we plant today, every plot of soil that we cultivate tomorrow, has the potential to open doors and enrich the lives of this generation and of every generation to come.

Concluding Remarks

Kenneth I. Shine, M.D.
President, Institute of Medicine

The presentations in this volume focus a good deal on interactions—interactions between health and fertility, between health and poverty, between mathematics and biology, and between education and science. They also highlight the relationships between the private and the public sectors, between profit and not-for-profit medicine, between the different health disciplines and the need to bring together—and, with hope, to educate together—a variety of providers who can function as an integrated team, and the need to reexamine the relationship between academic health centers and the communities in which they operate. As one step along that route, the Academy and the Institute have committed funds to bring representatives of 20 academic health centers and the local community schools with which they work to Washington for a five-day institute. At that time, we will introduce them to the new science standards, give them some experience with inquiry-based learning, and show them how to disseminate these science standards. If that works, we will do 20 more, and 20 more after that until all of the academic health centers have been involved in this process.

As our Nobel Laureate, Baruch Blumberg, has clearly articulated, we need to protect our capacity to generate new knowledge, and we need to understand the relationship between genetics and the environment, between behavior and illness. As someone who has spent his life studying virology and molecular biology, Dr. Blumberg is as eloquent as anyone could be with regard to the necessity to connect science to people all over the world. At the Institute, we will continue our commitment to improve and increase the acquisition and dissemination of scientific knowledge.

As IOM approaches its next 25 years, we will also continue our commitment to educate the new generations of health care providers. That includes a commitment to achieving balance in the health care workforce—an idea exemplified by the IOM's 1978 recommendation that 50 percent of all physicians graduating from medical school should be primary care providers. Now that managed care is expanding, we have an opportunity to focus not on the numbers, but on the quality and nature of primary care.

Finally, and most important, we must remain true to one of our most cogent values: our commitment to quality health care for all. In pursuit of that, we have established a roundtable on quality and have various other projects underway concerning the assessment of quality of care. This gives us a way to regularly evaluate the nature of the services that are provided to our citizens, their access to care, and the way in which the system responds to their needs. In addition, the Institute's special initiative on quality of care is designed to examine objectively and analytically the way in which the health care system operates so we can periodically tell the nation what is happening to quality of care and access around the country.

Although a good deal of the discussion at the symposium was focused on developments in the managed care environment as they relate to market share and for-profit activities, I should emphasize that IOM's commitment to *access to care for all* means that we will be steadfast in examining the way in which the changing health care scene provides care to all elements of society and to all of our citizens—the poor, the elderly, the indigent; those whose opportunities for care are limited by distance or by cost; and those who are put at risk because the consolidation under way in the health system may leave them on the outside looking into a health system that does not include them.

We here at the Institute are excited about the next 25 years. Let us hope that in 2020, our dream of quality health care for all Americans will have been realized.

Contributor's Biographies

BARUCH S. BLUMBERG, M.D., Ph.D., is Fox Chase Distinguished Scientist at the Fox Chase Cancer Center. Dr. Blumberg received his M.D. in 1951 from the College of Physicians and Surgeons of Columbia University and his Ph.D. in 1957 from Balliol College, Oxford University. He is a member of numerous professional and academic organizations, including both the National Academy of Sciences (1975) and the Institute of Medicine (senior member, 1982). He has received numerous honors and awards during his career, including the 1976 Nobel Prize in Physiology or Medicine.

ROBERT M. CAREY, M.D., is dean, James Carroll Flippin Professor of Medical Science, and professor of medicine at the University of Virginia School of Medicine. His research interests include fluid and electrolyte balance and hormonal control of blood pressure. Dr. Carey received his M.D. from Vanderbilt University in 1965. He is a member of a number of professional and honorific organizations, including the Institute of Medicine, to which he was elected in 1992.

LINCOLN C. CHEN, M.D., M.P.H., is Takemi Professor of International Health at Harvard University. He is also chairman of the Department of Population and International Health and director of Harvard's Center for Population and Developmental Studies. Dr. Chen received his M.D. from Harvard Medical School in 1968 and his M.P.H. from the Johns Hopkins School of Hygiene and Public Health in 1973. He has written extensively on health and development policy.

DON E. DETMER, M.D., is senior vice president, Louis Nurancy Professor of Health Sciences Policy, and professor of surgery at the University of Virginia. Dr. Detmer is also codirector of the Virginia Health Policy Research Center and maintains an active surgical practice. He received his M.D. from the University of Kansas in 1965. Dr. Detmer was elected a member of the Institute of Medicine in 1991 and is chair of the Institute's Board on Health Care Services.

JOHN M. EISENBERG, M.D., M.B.A., is chairman of the Department of Medicine, physician-in-chief, and Anton and Margaret Fuisz Professor of Medicine at Georgetown University Medical Center. Dr. Eisenberg received his M.D. from Washington University School of Medicine in 1972 and his M.B.A. from the Wharton School. Dr. Eisenberg is a member of numerous professional and honorific organizations, including the Institute of Medicine, to which he was elected in 1988.

RICHARD G. A. FEACHEM, CBE, PhD, DSc(Med), is Dean-Emeritus of the London School of Hygiene and Tropical Medicine and senior adviser in the Human Development Department at the World Bank. Dr. Feachem received his PhD in environmental health from the University of New South Wales in 1974 and his MD from the University of London in 1991. Among his many professional honors, Dr. Feachem was recently awarded a CBE by the Queen of England for his services in the field of international public health.

JEFF GOLDSMITH, Ph.D., is president of Health Futures, Inc., and a lecturer in the Department of Medicine of the Pritzker School of Medicine at the University of Chicago. He earned his Ph.D. in sociology from the University of Chicago in 1973. Dr. Goldsmith was a recipient of the Corning Award for excellence in health planning from the American Hospital Association's Society for Healthcare Planning in 1990 and has twice received the Dean Connely Award for best health care article (1985 and 1990) from the American College of Healthcare Executives.

PAUL F. GRINER, M.D., is vice president and director of the Center for the Assessment and Management of Change in Academic Medicine of the Association of American Medical Colleges. Dr. Griner received his M.D. in 1959 from the University of Rochester School of Medicine and Dentistry. He is a nationally recognized authority on medical decisionmaking and the delivery of health services. Dr. Griner was elected a member of the Institute of Medicine in 1986.

MICHAEL M. E. JOHNS, M.D., is professor of otolaryngology–head and neck surgery, vice president for medical affairs, and dean of the faculty at Johns Hopkins University School of Medicine. A specialist in the management of head and neck tumors, Dr. Johns is an internationally known cancer surgeon and is also well known for his studies of the effects and outcomes of a variety of cancer treatments—including surgery, radiation therapy, and chemotherapy. He received his M.D. from the University of Michigan Medical School. Dr. Johns was elected a member of the Institute of Medicine in 1993.

LAWRENCE S. LEWIN, M.B.A., is chairman and CEO of Lewin-VHI Health Group. Mr. Lewin has over 25 years of consulting experience in the health care industry, including multihospital systems, academic health centers, alternative delivery systems, and medical supply, device, and pharmaceutical manufacturers. He received his M.B.A. from the Harvard Business School. He was elected a member of the Institute of Medicine in 1984 and is currently a member of the Institute's governing Council.

EDWARD H. O'NEIL, M.P.A., Ph.D., is an associate professor of family and community medicine at the University of California, San Francisco, where he also serves as codirector of the Center for the Health Professions. Since 1989, Dr. O'Neil has also been the executive director of the Pew Health Professions Commission, which he started as a way of elevating health professional education and health work force issues in the debate on national health care reform.

JOHN E. PORTER, J.D., is serving his ninth term as the Republican congressman from the 10th District of Illinois. Congressman Porter is a senior member of the House Appropriations Committee and is chairman of the Labor, Health and Human Services, and Education Subcommittee. He also serves on the Foreign Operations Subcommittee and the Military Construction Subcommittee. He is founder and cochair of the Congressional Human Rights Caucus. Congressman Porter received his law degree from the University of Michigan.

DONNA E. SHALALA, Ph.D., is secretary of the U.S. Department of Health and Human Services. Prior to her service in the Clinton administration, Dr. Shalala was chancellor of the University of Wisconsin–Madison; as such, she was the first woman to head a Big Ten University. She has had a longtime interest in national science policy and has served on numerous committees and commissions related to science and technology policy. Dr. Shalala received her Ph.D. in 1970 from the Maxwell School of Citizenship and Public Affairs at Syracuse University.

KENNETH I. SHINE, M.D., is president of the Institute of Medicine and Professor of Medicine Emeritus at the University of California at Los Angeles (UCLA) School of Medicine. Both a cardiologist and a physiologist, his research interests include metabolic events in heart muscle, the relation of behavior to heart disease, and emergency medicine. He also participated in efforts to prove the value of cardiopulmonary resuscitation following a heart attack and in establishing the 911 emergency telephone number in Los Angeles. Dr. Shine received his A.B. and M.D. from Harvard University. He was elected a member of IOM in 1988.

DONALD E. WILSON, M.D., M.A.C.P., is dean of the School of Medicine, University of Maryland at Baltimore—the first African-American dean of a predominantly nonminority medical school. Previously he was professor and chair of the Department of Medicine at the State University of New York Health Science Center in Brooklyn and physician-in-chief of the University Hospital of Brooklyn and Kings County Hospital Center. Dr. Wilson received his M.D. from Tufts University. He is a member of numerous medical societies—including the Institute of Medicine, to which he was elected in 1993—and is a Master of the American College of Physicians.